SIZE DOESN'T MATTER

A BEGINNER'S GUIDE TO MAXIMIZING SMALL GARDEN SPACES FOR BIG HARVESTS

ERICA SUMMERS

I0459078

SIZE DOESN'T MATTER

A BEGINNER'S GUIDE TO MAXIMIZING SMALL GARDEN SPACES FOR BIG HARVESTS

Size Doesn't Matter:
A Beginner's Guide to Maximizing Small Garden Spaces for BIG Harvests
Copyright © 2026 by Erica Summers

Published by Rusty Ogre Publishing
www.rustyogrepublishing.com
Casper, WY, USA

Cover art by Erica Summers

ISBN: 9781962854559

MORE BY ERICA SUMMERS

The Rictus Grin and Other Tales of Insanity
Writhe
Bad God's Tower
Vanity Kills
Mantis
The Choice is Yours: Yakshar's Lost Treasure
The Choice is Yours: The Wishing Gem
The Choice is Yours: The Branson Ranch
The Choice is Yours: Escape From Sugarland
The Choice is Yours: The Australian Outback
Ensuring Your Place in Hell II
Price Slashers
Mourning Waffles *(As Trixie Fairdale)*
Mojitos & Murder *(As Trixie Fairdale)*
The Billionaire's Assistant *(As Odessa Alba)*
Rumspringa *(As Odessa Alba)*
Tangled Heirs *(As Odessa Alba)*
The Ugly Sweater Party *(As Odessa Alba)*

"Life's a garden, dig it.
You make it work for you.
You never give up, man.
That's my philosophy."

- Joe Dirt

TABLE OF CONTENTS

MY JOURNEY FROM GRAY THUMB TO GREEN THUMB

Many people walk past my tiny home garden, brimming with fresh fruits and veggies, and playful pollinators, and they say that they wish they had a garden like mine. I know this because my office window is within earshot of my garden on a highly-trafficked street, and I often have my window open. When I'm in the garden, many people walking by on their way to the beach at the end of the block tell me that they don't know how I do it, or they get too overwhelmed just thinking about starting one.

I always tell them that anyone can garden the way that I do (unless, of course, they live in Antarctica or Northern Canada, in which case, with such a small growing season, I imagine a lot of farming there has to take place indoors.) They usually laugh or balk at this and reply with something along the lines of, "I wouldn't know where to start."

1

I am writing this book for anyone who has ever thought or uttered that phrase. In this guide, I will show you step by step how to go from someone with a certified gray thumb or someone who has never gardened a day in their life to a gardener who has mastered their small space. Who knows, you may feel so confident after your first year or two that you decide to expand to a second raised bed or another area of the lawn or more hanging planters.

Every season, I give huge grocery bags full of produce to my boyfriend to take to his work to give out for free to his coworkers. Judging by the size of my zucchinis alone, most people think I operate a small farm. The truth is, I live in the suburbs in a tiny house by the beach in Connecticut. My yard is objectively tiny, curving around the house on a small fraction of an acre. Before that, I rented, and I mostly had to container plant on my patio. Even then, we were still giving tons of extra food away to the neighbors. Often, I would give them bags of hot peppers, and they could come back with some sort of family recipe hot sauce that they'd made with it and gift me a bottle as a thank you. It not only created goodwill between me and the people who lived around me, but it also opened us all up to a lot of conversation.

So, without knowing me, you might flip through this book and see all sorts of photos from my garden and think that I was born with some kind of green thumb or botanical gift.

You would be very, very wrong.

You see, growing up in Wyoming, my father had a garden. He would lose himself in it for hours at a time, this little six-foot by twelve-foot swath of our back yard. It was his time to decompress, to be alone with his thoughts, and to sweat like he stole something. We never helped him with it. He didn't want us to, and by God, we didn't volunteer.

2

Watching him through the window of the house, I would grimace. It didn't look fun. In fact, it looked like an exhaustive chore that he didn't need to be doing but somehow *wanted* to.

Years went by, and dad's gardens changed, his crops rotated, and he'd even sometimes go on and on about some new type of seed tape he just ordered from the Gurney's catalog and how he couldn't wait to plant it after the last frost.

Boring!

It wasn't until I turned nineteen and was diagnosed with cancer that everything sort of changed for me. Not because I had some strange change of heart or epiphany, mind you. No. I started to consider a garden because my oncologist suggested it. By the time they got me on chemo, I had amassed more than seventy tumors throughout my body, as the type of cancer I had starts in the lymphatic system. The good news was that, after a nearly year-long battle, I was finally off chemo and my scans were tumor-free. (Interesting side note, because I am a nerd: most of the main active ingredients in my type of chemotherapy were plant-based! So, in a way, I can say that plants saved my life.)

At the end of my chemo, my oncologist sat me down and told me that my immune system was now completely tanked. Something as simple as the flu could kill me. A regular cold could last for weeks instead of days. And I could pretty much guarantee that if anyone so much as sniffled in the room with me, I'd be ill within twenty-four hours. His advice to me:

Be dirty.

Basically, with me being severely immunocomprimised, he told me that allowing myself to get dirty regularly would help retrain my immune system over the following decades. If I got dirty enough on a regular basis and avoided antibiotics like the plague, he said that one day (far, far away) I could, in theory, be able to have a somewhat normal immune system again.

After a year or two of constant colds, strep throat, and a horrific flu season, I moved to Louisiana and got a house with a massive back yard and started giving the doctor's advice some real thought. After all, my dad had a garden. How hard could it be?

Turns out, *kind of hard*.

Don't worry, though. My book is about to save you years of trial and error and hundreds, if not *thousands*, of dollars in savings by learning from my mistakes early on in your gardening career.

You see, I didn't read any gardening books like you. I just sorta *went for it*. I bought a few packs of seeds and a few basic supplies, tore up a small square of my backyard to dedicate to an in-ground garden, and followed the directions on the seed packs. I watered it and waited. The house even came with a corrugated plastic and aluminum greenhouse, so I threw some soil from the ground in a few terracotta pots and dumped some seeds in there, too.

At first, things were going okay. Veggie seedlings started to sprout, and whatever my excitable Jack Russells didn't trample started to grow. A few weeks in, however, things started going south. Some started to wither. Some seemed stunted. Everything in the greenhouse baked to a crisp or was reduced to a few sad sprigs of its former self.

4

After a while, I didn't bother to weed the in-ground garden because everything was just a mess. I think I got a single handful of cherry tomatoes by the end of the season and nothing else. It just made me laugh to look at my pathetic bounty. *Great*. I was in the hole about a hundred and fifty bucks, and I grew about two dollars' worth of tomatoes at the local Rouses grocery.

The next year, I read a little online and got a wider variety of seeds to try, thinking at least this way I would be able to get *something* to grow.

Boy, was that not true at all. The second year was worse than the first. I'd sometimes sit at the base of my massive, towering Black Walnut tree in the middle of the back yard and stare at the patch of "garden" and wonder what I was doing wrong.

Turns out, I was literally sitting on the answer. The Black Walnut tree is kind of a fascinating tree. This one looked like it had been planted eighty-some years ago. It was massive with a trunk that could stop a heavy-duty pickup cold in its tracks without even shaking. The Black Walnut tree is interesting, though, because its root system does something very unique.

It poisons the ground around it. That's right. Not for animals, but for other plants. It is a process called allelopathy, a biological phenomenon that, in layman's terms, means the plant's roots either leach nutrients or secrete things that cause plants around it to decompose. It is the plant's way, in nature, for keeping a large area of ground for its own nutrient absorption so that it doesn't have to compete with other native or invasive plants for the fuel it needs. Some allelopathic plants store protective chemicals in their leaves, others in their bark. As the bark chips away and the leaves fall for the season, they decompose and become mulch and compost that poisons

5

anything around it.

Guess where my garden bed was, along with the patch of soil that I took for my potted greenhouse plants? That's right, right under a wide bough of that Black Walnut tree.

Not to get too nerdy on you, but the Black Walnut tree is extra tricky because it stores its allelopathic chemicals in its buds, roots, and nut hulls, too. And the toxic chemical it contains, Juglone, is most potent at the drip line, which was right over my garden.

Yikes! For nearly three years, I had just thought I was a terrible gardener with a gray thumb! All that time, I had told myself that it had to be something I was doing. Too much watering. Not enough watering. Wrong amount of fertilizer...

Nope! It was that gorgeous shade tree thwarting me!

The next year, I wised up. I created a raised bed and started planting in containers with soil from my FRONT yard.

What do you think happened?

You guessed it! Suddenly, my garden was blooming. I still remember how thrilled I felt that year when I came in with my bucket full of peppers at the end of the season, an amount that I now pull out of my garden WEEKLY in the summer.

The important thing that I want to impress upon you is that, no matter how grim it looked for me as a gardener, I didn't give up. Now that you will be armed with the knowledge in the pages ahead, I have nothing but confidence that your small garden will produce bigger and bigger results if you stick with it and apply these principles.

My next challenge as a gardener was when I moved to Connecticut. I gave up my owned home for a rental. We signed a two-year lease. The first thing my boyfriend said when we signed it was, "I'm sorry you won't be able to have a garden here. I know how much you love it."

My response?

"Watch me."

And I proceeded to grow buckets and buckets of produce right there at our rental property. I learned so many tricks for container gardening. I made plant hangers. I collected pots and rolling coolers during bulk trash days. And by the time our two-year lease was up and we moved across town, I had made friends with half the neighbors because of the garden, and I had inspired three of them to start their own.

Now, I own again, but I have NOT forgotten what it was like to have to maximize my tiny amount of garden space. The trick: companion planting. But there will be lots more on that later in the book.

Every time I am in my garden, I feel peace. I feel more in tune with myself and with nature. I work up a sweat, just like dad did, but now I understand that that sweat is full of pride. Pride that I know how to survive should the supermarkets all go empty. Pride that I grow enough to eat fresh food that isn't covered in toxic chemicals all summer long. Pride that my neighbors watch me digging that dirt, and within weeks, they're starting their own

7

gardens and telling me that mine inspired them to start. Proud that I can share fresh produce with my friends and my animals. (I feel like my snails love my zucchini almost as much as I do!)

In the coming pages, I will teach you how to start small and maximize what little space you're ready to dedicate to a garden, whether it is a single plastic pot, a four-foot raised bed, a small rolling grow-box, or an in-ground garden. Once you get the hang of it, you can keep expanding. If not out, then up!

Image 1 - A photo of me, proud, after one of my summer weekend harvests in Connecticut.

8

TOOLS: BUY ONE, DON'T <u>BE</u> ONE

Let's discuss some essential tools that you might need in your gardening journey. If you are gardening in a compact space, not all of these are necessary. Remember, you can always start with a couple of basics and expand every year, little by little. You do not need to spend a fortune on garden supplies to get a great yield of fruits and vegetables.

So what do you need? Well, here is a breakdown of some of the basics. You can decide what you need as you go. The only thing you truly need to start gardening is a seed, a container, and some dirt. The rest is all up to you!

GLOVES: Let's start off with a somewhat controversial accessory that can also be considered a tool: a good pair of *gloves*. While I personally don't like wearing them about ninety percent of the time in the garden, I would be remiss if I didn't state how important it is to protect your hands.

While it is fun to dig in the dirt (I swear I am genetically at least half Jack

Russell in my DNA with the way I like to play in the dirt), many beautiful things in your garden are full of thorns and barbs that want to give you so many piercings you look like that sad emo kid from junior high. *Whatever happened to him?*

If you are digging around in the ground for the first time, you never know what you will find. One day it could be a rock, another day it could be busted fragments of a beer bottle or a rusty nail. One time, digging in my garden, I found a set of dentures and a bathtub faucet. So, unless that tetanus shot is up to date, it is wise to protect your hands. There are also beneficial insects, such as spiders, that make their home in the dirt and don't take kindly to intruders. That is why a

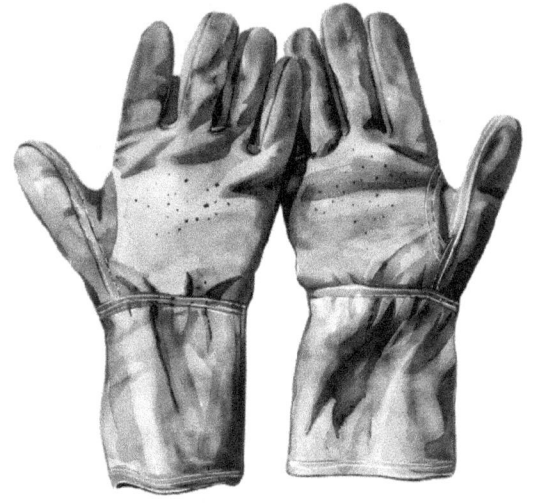

good, thick pair of gloves can be an absolute game-changer whether you're harvesting spiky pickling cucumbers off your vine, digging through brambles for those delicious raspberries, or pruning a fragrant rose bush.

Now, while all that is true, here comes the controversial part: There is actual scientific proof that direct contact with soil benefits overall health and wellness, strengthens immune systems, lowers stress, and improves mental well-being.

Handling soil with your bare hands, often called "a green prescription," puts one into direct contact with *Mycobacterium vaccae*, a bacterium that can activate neurons containing serotonin, the "happy hormone." This boost acts similarly to an antidepressant. Cancer patients reported a boost

10

in energy and mood when given a non-live version of this same bacterium. Pair that with the vigorous cardiovascular boost of gardening and the vitamin D infusion that being out in the sunshine and fresh air gives you, and you have a combo that will improve your mood while giving you a well-earned sense of accomplishment and filling your belly with nutritious food.

WATERING CAN: A watering can can be a handy tool, especially for a small garden or patio full of hanging planters or grow boxes. If you fill them up and let the water sit for 24 hours, chlorine will evaporate out of the water, making it an even gentler source of hydration for your plants. Just make sure to water the roots, not the leaves. If watering in the evening or on humid days when the plant leaves cannot dry  fully, watering the leaves can sometimes lead to issues like powdery mildew.

RAKE: A rake is a tined tool that allows you to gather leaves, debris, and detritus from your yard or garden space. If you only have an apartment patio, this will not apply to you. However, if you have a large yard, a rake can be your best friend. By gathering up clipped grass or fallen leaves, you can compile everything in either a compost bin or a compost pile, or in some cases, lay it directly on your garden bed. Also, if you are starting a new raised bed or getting ready to pot all new potted plants, you

11

can fill some of the bottom layers of your container or bed with grass clippings and old leaves, and throughout the season, they will break down and create compost that the fruit and vegetable roots can reach down and absorb. Think of it as a delayed fertilizer.

HAND SHOVEL: These are handy, no pun intended, for working in raised beds or planting seedlings. These are readily available at any retailer that carries garden supplies.

They do tend to fall apart after a few years if you are using them and straining them vigorously, like I do, but even the plastic ones will do you for a few seasons, and it sure can beat digging holes with your hands.

LAWNMOWER: If you have a lawn, there is a high likelihood that you have to mow it, too. Why not collect your grass while you're at it? Collecting your spent clippings can help add green matter to your compost bin or pile and, in turn, feed your garden, and therefore *you*!

HAND TILLER OR ELECTRIC TILLER: While this item is not absolutely essential, tillers do provide great results, especially in in-ground garden beds. I started out with a hand tiller, which is a four-pronged stick with a handle that can be jabbed into the soil and twisted. This aerates the soil and breaks

up compacted dirt so that roots can move freely and access the nutrients they need. It also helps with drainage. Once I upgraded to a larger garden, I started using an electric tiller, and they are amazing. They make quick work of larger garden spaces. As I have mentioned before, my garden is only about the length and width of a standard driveway, but that tiller can chew up the top six inches of dirt in a matter of seconds. It makes the top few inches of soil soft and airy, and your plants will thrive if tilled before each season. This is not necessary for potted plants and window planters.

HAND CLIPPERS, PRUNERS, SHEARS, OR SNIPS: These are inexpensive

and essential items for any vegetable garden or fruit tree owner. These are typically used for harvesting tomatoes and herbs, pruning roses or blueberry bushes, cutting back annuals and perennials in the fall, and even deadheading flowers. You will want to sterilize them in between uses to keep them from spreading any illnesses between plants, but this can be as simple as a pair of household scissors.

13

LOPPERS: Loppers are heavy-duty pruners with long handles made for chopping limbs off small trees or shaping shrubs.

GARDEN HOE: A hoe has a flattened metal head with a bent neck that is attached to a pole. Gardeners use these to loosen soil and dislodge weeds.

IN-ROW WEEDERS: An in-row weeder looks like two bent rake heads on one stick, and when dragged through plant rows, the flexible prongs lightly stir the soil surface and can thin plants and eliminate some annual weed seedlings. These can be particularly good for onion beds.

CANVAS SOAKER HOSE: These are great for small gardens. Canvas soaker hoses provide a form of irrigation that cuts down on powdery mildew and fungal infections because they snake through the bases of the plants and dribble water when the spigot is turned on. Because the leaves are not sprayed with water, they do not stay moist, so the water soaks down into the roots but doesn't ever get the chance to create moldy environments for the leaves. When paired with a watering timer, your irrigation can be fully automatic. If you plan to have a single raised garden bed, I cannot recommend the canvas soaker hose and timer enough because if you set them up correctly, you will not have to water your garden at all, and you will be able to reap all of the same benefits of a gardener who is out watering plants every other day.

Many brands of soaker hoses can be connected together for longer strips of irrigation however, there is a tipping point. Sometimes, when you connect too many of them at once, you will have heavy irrigation near the spigot, and not enough water will reach the end of the hose in some cases. If you plant your thirstiest plants near the *start* of the soaker hose, and the plants that like the driest legs at the end, you can still create a perfect irrigation environment for your plants.

GARDEN HOSE: I personally love the rubber, retractable hoses that are now on the market. For many years, I struggled with stiff, kinky hoses, and while they do get the job done, they can be difficult to put away. Either way, as long as you have a method of getting water to your plants, that is what is important. By having a multi-setting spray nozzle attached to the end of it, you can water different crops in a way that suits them. For example, I typically water my 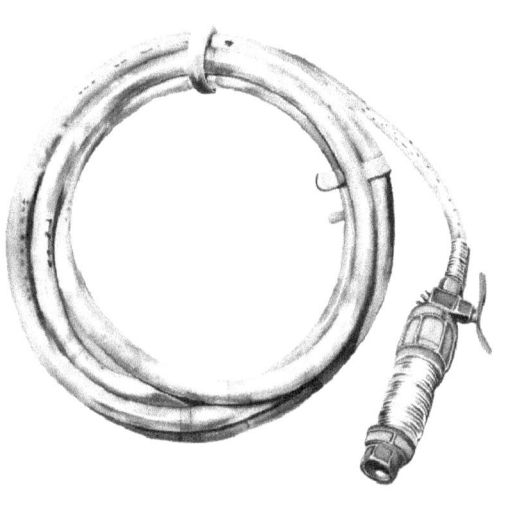 strawberries with a rain-type setting to make sure that they get water uniformly and that the roots can spread out in many directions. However, I typically water my tomato and pumpkin plants with a jet-type setting to keep any water from hitting their leaves.

WHEELBARROWS AND GARDEN WAGONS: These are typically only used for larger gardens. Spare yourself the expense if your garden is very small and you don't really have a room to expand. However, if you have a larger garden area, these can be very handy tools to save your back from all the heavy lifting of bags of dirt, compost, and mulch, should you choose to buy any.

PLANTER FLATS/SEEDLING FLATS:
These are small plastic containers with uniform cells used for starting seeds either indoors or outside. These typically come with a clear plastic lid and a black plastic water catch tray. These are only necessary if you decide to start your own seeds before your grow season has officially started.

HAT, BANDANNA, OR SUNSCREEN:
One of the most magical properties of any garden is the ability to lose complete hours out there, playing in the dirt. The sun's rays can be brutal. Even during the heat of the summer, I typically only weed or do anything more involved than a walk-through three times a week. On those three days, I typically alternate between a sweatband or bandanna one day, a big floppy hat another, and then sunglasses and sunscreen on the third, depending on the weather. They all work. You just need something to keep the sun and sweat out of your eyes. Gardening can get pretty vigorous!

SEED ORGANIZER:
These are optional (you can always keep your leftover seeds in their packets and tuck them in a plastic baggie to keep the moisture out), but man-oh-man are they handy. Seed organizers come in a variety of

16

shapes and sizes.

My personal favorite is the one that contains a bunch of empty individual airtight containers that look like TicTac cases. They store thousands of seeds, and they're so easy to carry around in the garden. They take up less space than the more traditional organizers, too. However, some people find the tiny labels tedious or annoying. Either way, if you have the money and space, a seed organizer will save you time and effort in hunting things down or from accidental spills.

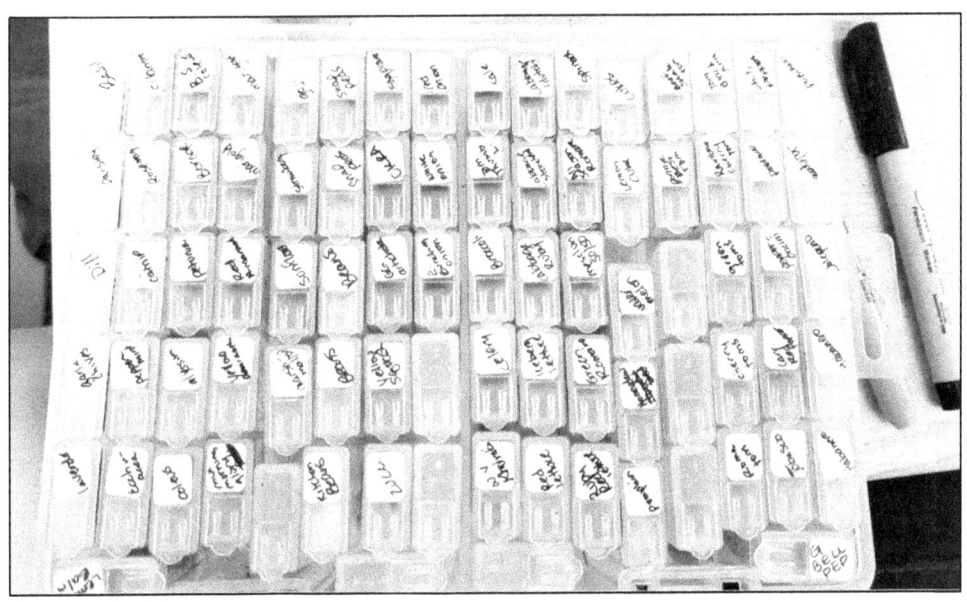

FOAM KNEE PAD: These things are great. They're super cheap, super lightweight, and they really save your knees when you're transplanting seedlings or weeding the garden. They're a narrow foam strip with a wide slot cut out at one end that acts as a handle. You don't need to spend more than a couple of bucks on one.

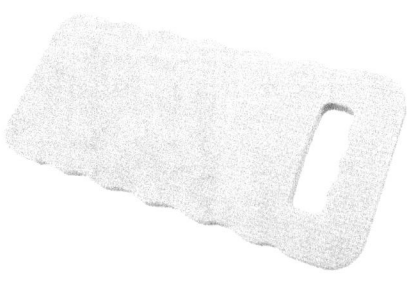

17

These aren't an absolute necessity, but they sure are a godsend when you're kneeling on hot, dry soil or mulch.

SHOWER CURTAIN: These are handy (but totally optional) when creating a compost pile. They lock in moisture and heat and deter vermin. Cover your compost and add a few bricks on the sides to keep it from flying away.

SOIL METER: Most soil meters measure pH, light levels, and moisture content through dual probes. Simply jam the probes into your soil and select what you want to meter. These are a great way to gauge whether you need to water again, whether a plant is receiving adequate light, or whether your soil's pH is out of whack.

I highly recommend these for outdoor plants as well as indoor ones. These have saved me from over-watering countless times!

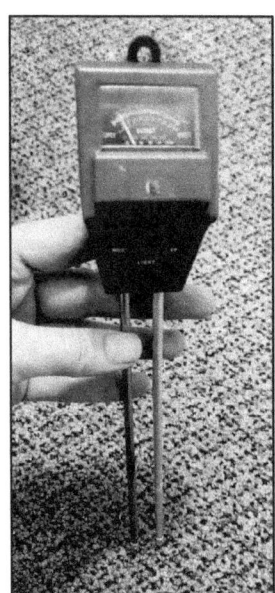

Image 2 - An example of an inexpensive soil meter.

TIPS FOR TOOLS

You can protect garden tool handles by spraying the wood with a thin coat of hairspray. This can preserve the wood and protect it from the elements

You can disinfect your tools after you have used them on your garden with a mixture of diluted bleach and water or hand sanitizer. These kill bacteria and fungi.

Keeping the metal from your hand shovel submerged in a pail of sand is a handy way to protect the metal while keeping the edges sharp.

DO YOU RENT OR OWN?

Do you rent or own your home? Do you share an apartment with others and only have a window? Do you live in a high-rise and only have a balcony? Do you rent a home with a big yard but are worried about jeopardizing your security deposit? Or do you have a whole yard to get creative at a place you own?

No matter which of these applies to you, there are options! You can grow a compact garden no matter your residence's limitations.

When I started my gardening journey, I owned my own home. This is certainly the easiest of the above situations because you have total freedom to do whatever you want. I could expand freely, remove tree limbs to give my garden more light, etc. When I moved to Connecticut, my boyfriend and I decided to rent for a little bit so that we could shop around and buy the house that was right for us. The only suitable place that we found at that time required a two-year lease. Being that this was during the beginning of the pandemic and places were hard to find, we jumped on it. My boyfriend, Dave, thought that we wouldn't be able to have a garden in a rental, but I assured him that I am stubborn and reminded him that *where there's a will, there's a way.*

Immediately, I got to work creating a spacious garden. I asked the landlord for permission to do some branch maintenance on the overhead trees, and he approved it. By trimming some of the branches overhead, I was able to create a nice dappled strip of light so that my plants wouldn't fry in the summer sun.

We put up a temporary fence for our two Jack Russells made of rabbit wire zip-tied to metal pound-in four-foot stakes placed every few feet. Since I didn't want the pups trampling the seedlings, I used the spare rabbit wire and stakes to create a three-foot rectangle bordering one side of the dog's fence.

Image 3 - An aerial view of the rental's yard after we installed temporary fencing. On the right side, next to the composter, you can see the tilled 3-4 foot wide garden strip.

21

Now, I realize that if you have a one-year lease or a landlord that does monthly check-ins, or there is something in your lease that prohibits this sort of thing, this method of in-ground gardening may not be an option for you. Because I had a two-year lease, I knew the first year, I could grow in-ground where plants often have a cache of nutrients built up in the soil to ingest.

I also knew that the second year, I would have to let the grass grow over and keep everything inside that fenced-in area in pots. At one point, I even had a few shelves I made from spare wood laid across a few cinder blocks and rocks around the property so that even the containers wouldn't leave our grass polka-dotted when it was time to get our deposit back.

Image 4 - Aerial view of the garden strip with some of the pots consolidated into one spot. I did this to keep weeds from forming while I waited for the last frost.

The fence I built the first year kept the deer and many of the groundhogs and rabbits away from all of our fruits and veggies, even when they were in the containers.

I knew that once we purchased a home, I'd be able to in-ground garden again. That second year, I learned how to get *very* creative with my container gardening. That's when I also learned about that absolutely

22

magical combination of words:

Companion planting.

Through this method, I learned how to plant in a way that would maximize the tiny amount of container space. I learned to pair or group plants together that had beneficial, symbiotic qualities with one another, too. But more on that later, I promise.

I also learned how to select vegetables and fruit and fruit trees that would grow well in the partial shade of my raised, roofed back patio. Since I couldn't put anything in the ground that second year, I learned how to make plant hangers out of yarn. I used felt grow bags and window planters. I used plastic pots and bins, and drilled drainage holes in trash cans and cheese puff containers, and planted in pretty much anything that would hold dirt. I used a screw gun with a wide drill bit to drill drainage holes into anything and everything I thought was worth a shot.

Did it look a little janky? Yes.

Did it work?

Absolutely.

That year, all my garden cost me was a few packs of seeds, some water, and my time. I had containers on every porch step, plants hanging from D-rings, nails, and balcony railings. I created a patio jungle that was both lovely to enjoy a morning coffee or afternoon glass of wine in and in turn, by the end of the second year's harvest, I was pulling literal storage buckets of produce out of that container garden.

Image 5 - A photo of large potted plants on each patio step.

I say this not to brag but to illustrate my point that vegetables and fruit can be grown even without having a dedicated in-ground garden.

Now, there are some caveats to growing this way. Some of the cons of growing like this are that the plants in pots and hangers might need to be watered more frequently, as they can't send roots down to the moist soil below on dry days.

Conversely, it can be a little easier to drown or water-log your plants if there aren't enough drainage holes or if your soil is packed down so tightly that it can't drain. The roots of your plants can suffocate in too much water. This can stunt its growth, force it to struggle with fruit production, or kill the plant altogether. This is easily cured by mixing in some amendments with your soil prior to transplanting seedlings or direct-sowing seeds. For this, I recommend Perlite or compost, both of which I will cover in a later chapter. Alternatively, larger or more holes can be drilled into the base of your pot before (or even after) planting.

A great trick, if you are having to use pots like this, is to hang other fruits and vegetables in planters via a hanger, nail, or shepherd's hook a few feet over each potted plant. That way, when you water the hanging plant above, the one below will get the hydrating run-off instead of your lawn. This can also provide the added benefit of providing a few minutes of partial shade around noon to plants that prefer that sort of thing (like squash and zucchini, lettuce, carrots, etc.)

Another great option for renters or balcony planters is the self-watering grow box. We will discuss those more in the next chapter on containers.

If you own your home, you have a lot more freedom to roam. I encourage you to start small and expand yearly. So many gardeners start out trying to do it all, and many take the failures pretty hard. But here's the deal: You're going to have failures no matter how experienced you are. Half of gardening is problem-solving and battling pests and elements. Problems always arise. Some issues might be your fault because you went out of town and your friend forgot to water. Some will be completely out of your control because a hurricane or storm hung around a while and kept conditions so wet that your vining crops ended up with a bad case of powdery mildew or snapped tomato stems.

Starting on the smaller side keeps the stakes lower and can keep you from getting overwhelmed when problems arise.

GARDEN MAPPING & PRE-PLANNING

You might have seen the heading to this chapter and sighed while thinking, *Is planning absolutely necessary?* My answer might surprise you.

Absolutely not.

However, in my opinion, planning can make a world of difference if you are actually planning to companion plant fruits and vegetables. However, if you are more chaos-oriented, you might also really enjoy something called chaos gardening.

I will go over both of these methods and give you some great examples of how I do mine with some paper, a pen, some sticks, and a roll of cheap paracord.

CHAOS GARDENING

Let's start with chaos gardening since it is the easiest method and requires very little preparation.

Chaos gardening is the almost completely indiscriminate sowing of seeds, usually in bulk, whether in a container, raised bed, or in-ground. It is a minimal-effort method of planting that allows a gardener to mix a variety of seeds and scatter them in an area. Then, all they have to do is water and occasionally weed. This is a favorite among many beginners because it's inexpensive, takes almost zero pre-planning, and feels spontaneous.

Chaos gardening is mostly used for pollinator flowers or herbs. Often, you will see big pre-mixed bags of seeds at the store that have already been combined and are ready to spread on your fresh-tilled or roughed-up ground.

There are a few minor downsides to this type of gardening:

One is that there are often heavily reseeding annuals in these pre-mixed bags. You might find that the morning glories that bloom along your fence one year are the bane of your existence three years later when they are strangling your fruit tree like a pair of grabby hands. You haven't planted them in years, but now they just keep coming back like a pretty weed.

Another downside is that when someone walks by and says, "Wow, that white flower there is stunning! What is it? I'd like to grow some in my yard," all you can really do is shrug and smile unless you really know your flowers. Trust me, unless you live in a sparsely populated area, this actually happens a *lot* more than you might think.

Another small downside is that you do not have control over what you planted. You might have things that are beautiful, but they are planted far too close to one another by the time they start to mature and bloom, and they can choke out the nutrients, water, or light the smaller plant needs.

One thing I don't like about chaos gardening is that it feels like only a tiny fraction of the seeds I toss at the ground actually sprout. I have a hard time distinguishing weeds from unknown flower seedlings, too, sometimes, so I'm sure I've probably pulled many that would have been beautiful just because I didn't know what I was looking at.

There are many positives to chaos gardening, though. One of which is the ease of it all. There is almost none of the labor that'll make you sweat involved, so this can be a great method for someone with an impaired range of motion or an injury.

It is also a fabulous way to fill a pot or container that isn't food-grade with something that will bring a boatload of pollinators to your garden. They're great for hellstrips, too, but I talk about that in greater detail later in the book.

The trick to chaos gardening is sowing seeds of plants that will not only survive, but thrive, in the conditions in which you're laying them down. For example, if you want to brighten up the ring around your maple tree with some great pollinators, you shouldn't cast down ones that like full sun. After all, they will be under the shade of that maple for a massive chunk of the day. Likewise, if you want to plant things that need partial shade, they will fry like an over-cooked onion ring in that cute little full-sun patch by your mailbox.

I will say, I have chaos gardened before in pots and hellstrips, and it is quite fun. If you are not sure you want to commit to a full-blown compact garden, give it a try for sure. But I suspect, based on the title of this book, that you have greater aspirations.

So let's talk about planning and how it actually isn't as tedious as it sounds.

PLANNING YOUR GARDEN

As you might have already guessed, I am a big fan of actually planning out a garden. Partly because I live in a place where it snows for a bit of the year, and while I have that itch to garden, it gives me some time to plot and plan ways to really get the most out of my small garden space. I give away enough produce that people often think I live on a farm, but as you can see from some of these photos, my garden is just a little bigger than a single,car driveway. This should be a testament to mapping out a garden and pairing companion plants to get the most out of your small space.

First, every January or so, I start with a list. I poll my boyfriend and his coworkers to see if there is anything that was a big hit last year, or find out if anyone has any requests. I, personally, love the act of watching something grow. I feel pride knowing that I created it with my hard work, and I feel a real sense of joy when I am eating fresh food and still have enough to give away some to neighbors, friends, and coworkers. It builds a sense of community and often even inspires others to do the same.

Can you imagine how great this world would be if a lot more people had their own gardens? We wouldn't be facing such a crisis with all of these spray-on or sprinkle-on bug killers that are decimating the pollinator population.

This world needs a lot more flowers and gardens than manicured lawns, in my opinion.

Image 6 - Street view of the uniform rows of the in-ground garden seasons.

So how does one actually plan out a garden? It's easy and surprisingly fun, especially if you are one of those nerds like me who like a good mind puzzle every once in a while.

Some people go for nice neat rows of things, like the street view of one of my seasonal gardens, which I will admit can look very aesthetically pleasing. Nothing says "I know what I'm doing" quite like long uniform rows of things.

But that is not the only way. Some people plant just a few of each type of plant they want to attempt, and there isn't enough for a row. During my boyfriend's first year of growing seedlings, he thought about the plants he'd need for certain dishes. He planted Roma tomatoes, Black Beauty eggplants, and tons of hearty basil. Flashing forward to that fall, I must say his eggplant parm was on-point! I, myself, like to experiment and grow new things every year to determine the things that I really enjoy growing or things that do very well with my soil type. This is exactly how I discovered my love of growing lemon cucumbers and learned the funky process of how banana melons grow.

If you are starting with a small raised bed or garden planter, I advise you to take full use of the *Companion Planting* chapter later on, as this will help you get the most out of your compact garden.

Image 7 - Aerial view of the in-ground garden after I tilled and pinned down strips of fabric weed barrier.

Once in a while, I wait until spring. Then, I map it after I till the ground the first time and lay down some walking paths so I know how many rows I can fit. However, most years I start mapping my garden in the winter, before I've bought any new seeds. When it is cold out, I'm aching to plant again. That's when I sit down with a pen and paper, I make a wish list of everything I want to grow in the coming season, and then I get to sketching.

31

When mapping out my garden, I like to crudely sketch it. It doesn't need to be anything fancy or professional looking. It is just a guideline so that you know how many seedlings to start in your flats, what will be direct-sown after the last frost, and so on.

When sketching it out, I can see what I can plant more (or less) of, or if I need to dial back on something completely. It also gives me time to look at things that might not work together well before they are in the ground, vying for the same nutrients or spots of sun. This works especially well if you only have the space for a four-foot raised bed, a grow box, or even a few hanging planters.

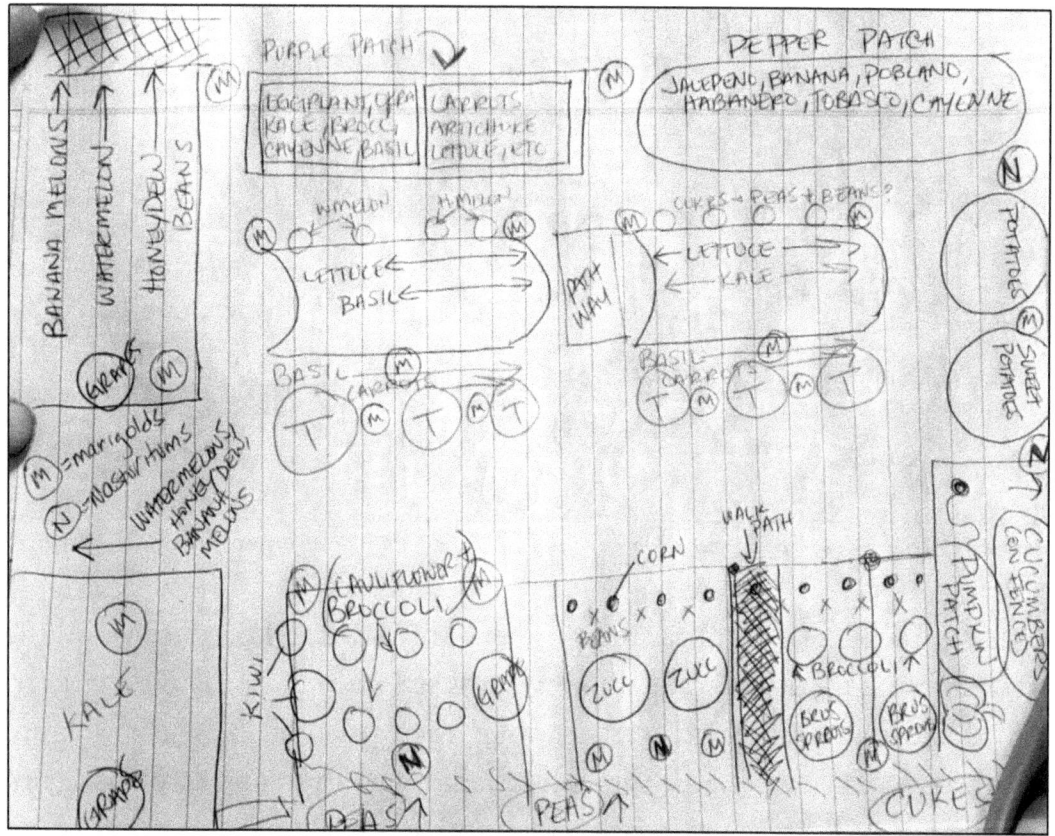

Image 8 - One of my winter garden plans for what to companion plant the following spring.

32

Image 9 - A view of my in-ground garden based on the sketch above from months prior. This image includes my string guidelines so that I knew where to walk and where to transplant my seedlings after the last frost. Note: While this wide-angle picture makes the space look fairly sizeable, it is only just a bit larger than a grocery store parking space.

There are a lot of factors to take into account when mapping, such as *which side faces south*? Assuming that you live anywhere in the Northern Hemisphere, the south side of your garden will receive the most sun. (Conversely, if you live in the southern hemisphere, the north side of your garden would receive the most sun) So if you live in the northern hemisphere (Like North America, Europe, Asia, etc), it is important to plant the taller plants on the north side of your garden. Planting them on the south side will ensure that everything will get a lot more shade. I always take this into account and try to group the smaller plants like root crops, peppers, lettuces, zucchinis, and melons on the southern half of my garden while leaving the northern side for my giant sunflowers, tomato vines, okra, and anything that requires a trellis. However, there are some combinations of crops (such as "the three sisters" that I mention later in this book) where planting tall plants on the south side might actually yield better results.

33

However, as a *very* general rule of thumb, you want your small plants on the southern side.

Next, you want to look at your wish list and find companion plants that work together harmoniously, and then mark some space for them on your paper. Depending on your grow area or what you desire to harvest, your plans might look wildly different than mine. This is a time to be creative and try new things. If you want ten varieties of tomato in your garden and nothing else, you can do it! There is no rule or law that says what you can and can't plant, with the exception of marijuana, which is only legal to grow in some states. Also, some states have regulations for the number of tobacco plants you can grow for personal use as well, so make sure you do a little digging if you plan to grow those in your home garden, as you will need to check your state's rules and regulations beforehand. (That said, I have legally grown both of the aforementioned plants, and they were some of the most gorgeous plants that ever came out of my garden.)

Image 10 - My back garden second-year expansion. This was all just a shoddy slab of asphalt when we bought the home. The grass is AstroTurf, and with the grapevine shade overhead, it is a great place to do yoga.

34

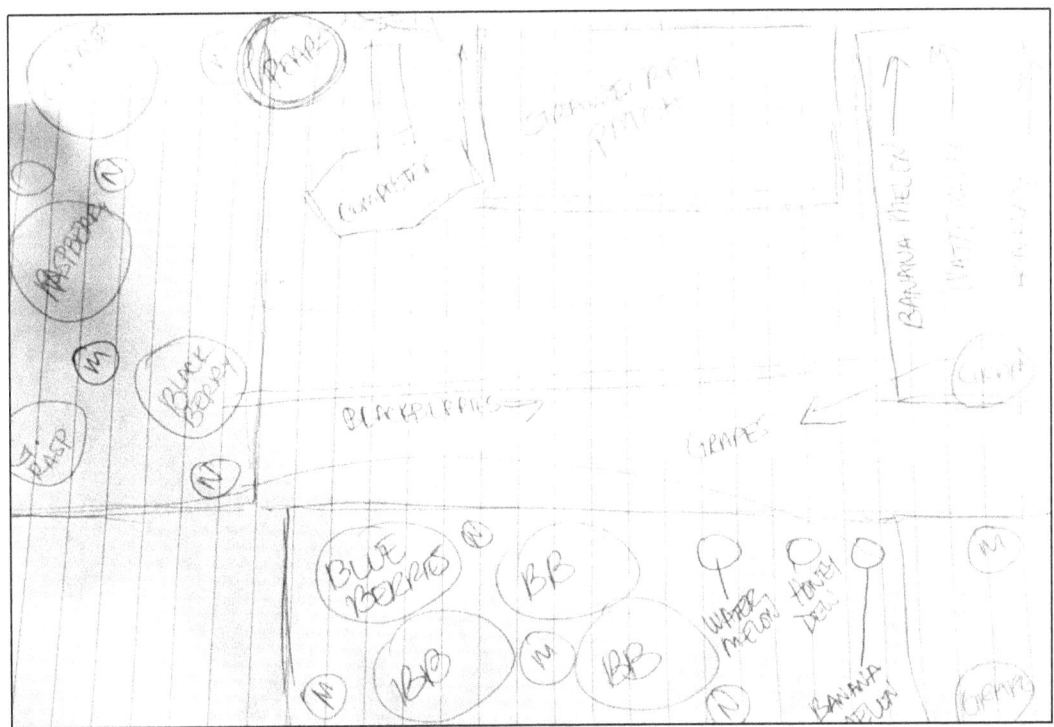

Image 11 - A sketch of the garden shown in the image above made months before.

After I map out my garden, I have started doing string guidelines with sticks and paracord to map out the areas. Before, I just sort of guessed. This led to me trampling to death a few budding seedlings that I direct-sowed into the dirt. I think I even pulled some things I planted, thinking they were weeds, because I couldn't remember how far over I had planted things. With the string guideline (shown in one of the earlier images of this chapter), I know what seedlings go where once they harden off. I also know where to direct-sow things like corn right into the soil. Most importantly, I know where to walk without crushing things I just planted.

All you need is paper, a pen, and a little imagination to design your perfect beginner garden. I encourage you to give this method a try after reading the *Companion Planting* chapter!

35

Image 12 - A picture of the same garden from the prior map just a few weeks later taken from another corner.

GARDEN TYPES

In this chapter, we will discuss various garden types to find out which one (or which combination) will work best for the space in which you intend to garden.

RAISED BEDS

Raised beds are elevated garden beds that can be erected on top of weed or mesh barriers, concrete, grass, or bare dirt. These give you a lot of flexibility with your soil quality as well as the texture and compaction.

Raised beds can come in a variety of materials, from prefabricated aluminum that can be shipped flat right to your door and assembled quickly to heavy-duty wooden ones. (Those are simple to make out of a couple of L-shaped brackets and some planks from your local hardware store.)

Personally, I prefer the ready-to-assemble aluminum ones. They are not only extremely affordable nowadays, but they can be ordered with ease. If you like to think outside the box like me, you can even erect them into custom shapes instead of the traditional rectangle or oval that they come in. For example, with three rectangular raised beds, I was able to assemble them together as one unit in the corner of my back lot and build one in the shape of an "L." This flexibility allowed me to create an entire corner of my garden on top of some old asphalt that was once used as a makeshift driveway.

When I moved into my home, the broken asphalt was unsightly, and my boyfriend thought nothing could be done with that section of the yard. I assured him that it was a massive area that would be perfect for an above-ground garden. For less than $80, I was able to purchase all of the aluminum beds and have them shipped to our home, where I assembled them in under an hour by myself. They're fantastic.

Image 13 - Two halves of another aluminum raised bed I assembled in less than half an hour.

The trick to starting a raised bed on concrete or asphalt is to layer in a few

38

inches of cardboard first. (Make sure you take any tape off the cardboard and that it isn't covered in ink.) I flattened boxes and made a nice, thick layer of cardboard. Then, I layered in two inches of shredded paper and two or three inches of dirt. Then, I added some of my homemade compost and a few inches of dried leaves and grass clippings. I filled the rest in with dirt all the way to the brim. I call this the *lasagna process*.

As it starts to rain, and the worms and microbes start to devour the paper and organic materials, the cardboard and shredded paper flatten, and the top layer of dirt will sink slowly. I skipped that step and watered it thoroughly. In a day or two, I added even more dirt to top it off.

Image 14 - Once the raised bed was in place on my patch of asphalt, I filled it with a layer of cardboard, shredded paper, and branches I pruned off a nearby mulberry tree.

39

Image 15 - This is another raised bed that I layered using the same lasagna process with cardboard, sticks, shredded paper, and dirt. After I reached the halfway point shown in the photo, I filled the rest to the brim with soil and more compost.

I had a beautiful, raised bed that beckoned earthworms from the nearby dirt. The next spring, I filled it full of fruit and vegetable plants and had so many cucumbers and banana melons that I was giving bags of them away!

Every year, I add fresh compost,, a little more dirt, and mulch, and year after year it has produced gorgeous fruits and vegetables.

Some of the cons for raised bed growing are that the soil can get depleted of its nutrients quicker than in-ground beds. It is very important to add fresh compost every year and fertilize a bit during the vigorous growing months so that your plants are getting everything they would normally be getting from the ground.

In the past, I have also assembled *wooden* raised beds that work well, especially when placed on top of bare soil. The nice thing about those is that, like the aluminum raised beds, you are able to weed-eat around them to make the grass around the bed look tidy, so your plants really stand out.

Image 16 - A very cheap, slightly raised bed that I made with three pieces of scrap wood lying around in my shed from other projects. I planted all purple items in it and called it my eight-foot "Purple Patch." You can see this in the sketch in Image 8.

I should note that you should use *untreated wood* for this project, even though this, in itself, shortens the life of the wood. Treated wood, as well as varnish or paint, on the *inside* of the bed, allegedly leaches small amounts of toxins into your fruits and veggies. Using bare, untreated wood, however, works great. You can make a nice, simple eight-foot-long raised bed for less than $20 if you have a screw gun, a couple of fifty-cent brackets, and about thirty minutes of time.

This year, I created my eight-foot-long "purple bed" (a raised wooden garden bed dedicated solely to fun and funky purple fruits and veggie varieties shown in Image 16) with about twelve screws, three pieces of scrap wood, six metal L-shaped brackets, and one single cut with a saw. It worked like a charm and cost me less than a cup of Starbucks coffee.

IN-GROUND

In-ground gardening can produce some of the absolute best results and the most incredible fruit and veggie hauls week after week during your growing season. That said, they might not be an option if you rent, and they do have a few downsides.

In-ground beds are phenomenal if you can spare the space. They are easy to water, and they can suck down a lot of excess rain during long rainy periods that might otherwise flood raised beds and containers. Another huge pro for this garden type is that the plants have access to a variety of nutrients and minerals that a container or raised bed might not.

There are a few things you should know before choosing these, though. One thing to note is that they require a substantial amount of weeding compared to the other garden types. This is because grass likes to spread, and weed seeds scatter in the wind and embed in the ground near your plants. This can be a turn-off for a lot of beginner gardeners because it takes a substantial time commitment each season to keep these weeds at bay so that the plants you don't want aren't choking out or stealing nutrients, or casting shade on all of the plants that you *do* want.

That said, there is no such thing as a weed in most gardeners' eyes. A weed is simply something you don't want growing in a certain spot, usually something you didn't plant. Consider for a moment that my maple tree sheds thousands of helicopter-like seed pods every year that spring up all over my yard. To me, that is a weed because I didn't plant it, and my tiny yard can't sustain one hundred giant maple trees. Now, somewhere in the world right now, someone is spending a *lot* of money on a three-year-old

42

sugar maple to plant in their yard. To them, it isn't a weed, but an ornamental addition to their landscaping. To me, they are a minor nuisance that needs to be plucked regularly.

There are lots of things that most people pull as weeds that also have some major health benefits. Take purslane, for example. This is something that gardeners everywhere pluck like mad and throw away. It comes back every year. But most gardeners don't know that this "weed" is rich in omega-3 fatty acids, potassium, magnesium, and calcium. It contains vitamins A, C, and E, all of which boost the human immune system. It also contains beta-carotene and has been said to reduce the risk of heart disease and even cancer. It can reduce your diabetes risk, too. It can help regulate blood sugar and has anti-inflammatory properties and fiber, which promotes digestive health. All of that aside, it tastes pretty good, too. Whether eaten raw in a salad, sauteed, or boiled, it's not bad.

I could go on all day about purslane and a variety of other beneficial "weeds," but the point I'm trying to make is that they shouldn't be vilified. Some should be embraced!

Image 17 - Purslane

43

Image 18 - An in-ground bed can be extremely rewarding.

Another tricky thing about an in-ground bed is that you have the ability to walk on it, compacting the soil so that the roots have to work much harder to spread. This is why establishing walkways where you can still access everything is crucial. As you can see in the image above, I have a few long window planters mixed in throughout the area.

I did this strategically. They are full of flowers that entice more bees into the yard to pollinate my flowering crops (like zucchini, pumpkin, eggplant, tomatoes, etc) while also adding a nice pop of color for anyone walking by. The best part about those planters, though, is that they are movable. When I need to harvest the zucchinis or broccoli (at the bottom of the photo), I only have to slide those window planters out into the aisle, and then I can reach the back row without trampling my crops. Then, when I'm done, I simply slide them back in place. They prevent weeds, make weeding the crop squares much easier due to the access, and they function in a wonderful way because of the flowers they contain!

44

Image 19 - (Top) My freshly-tilled garden area in April 2023.
(Bottom) The same garden in June, just two months later.

CONTAINERS

Containers can make a wonderful compact garden because they are literally everywhere. Beside being readily available, there are some real upsides to container planting.

There are many types of containers that you can garden successfully in.

A NOTE ABOUT MICROPLASTICS AND NANOPLASTICS:
Before we get into the ins and outs of all of the different types of containers available to gardeners, there is something I feel I should bring to your attention before we get started: microplastics and nanoplastics (or NPs).

Some plastic pots, bags, and containers are not considered "food grade." This is something that I learned about several years into my gardening journey. I found out that not all of the pots made of plastic that I was growing in were considered "food grade."

I also found out the degradation of old plastics and synthetic fabrics can lead to a buildup of micro-plastics in in-ground garden soil, too.

According to the *World Economic Forum*, humans inhale roughly 68,000 microplastic particles each day. Microplastics have now even been detected

46

in human lung tissue as well as the liver, blood, lower extremity joints, and even semen. *Why am I telling you this*? Because it is worth noting, especially if you are particularly health-conscious or immunocomprimised like I am.

Nanoplastics impact gardening because when plastics degrade, break, or disintegrate into your soil, they can be absorbed through plant roots and, in turn, can begin to affect the quality of your plants. When a plant has absorbed a lot of these microscopic nanoplastics, its water and nutrient absorption abilities can be slightly hindered. This can also increase the plant's ingestion of pollutants (such as arsenic and heavy metals).

I did a lot of digging on this particular subject. The consensus among medical professionals is simply, "Don't do it" (the same thing they say about soda, salt, and your daily deodorant) when it comes to using non-food-grade plastic containers. Most sites and sources could not prove that this is actually enough of a problem for the household gardener to stress about, but if the thought of microplastics is concerning to you, I urge you to do a little digging of your own before deciding. Better safe than sorry.

Personally, most of my gardener friends will use just about *any* container they like the look and size of. Many swear by their plastic pots and containers. While I usually only use my more questionable plastic pots for flowers these days, I will say that I used plastic containers for growing my own food and fruit trees for nearly a decade (unknowingly) without a problem.

With this section, I simply want to inform you that these things *exist* so that you can decide whether planting your fruit trees or veggies in a plastic container is something that you want.

So... are *all* plastic containers a terrible idea to use in your garden? No. We consume microplastics every time we drink a bottle of water or eat Chinese food out of one of those to-go containers. It is in the trash that the wind sweeps into our yard. It is in the things that sat on our grass before we ever moved in.

You might ask, is there a way to know if some plastics are safer to garden in than others?

I'm happy to report that there is!

Food-grade plastics are manufactured specifically to be safe for contact with... You guessed it: *Your food*. It is usually free of dyes and additives. To find a container that is deemed food-safe by the professionals, you can look for containers marked with any of the following:

- PP (polypropylene)
- LDPE (low-density polyethylene)
- HDPE (high-density polyethylene)
- PETE
- Food grade/ Food safe
- BPA-free

The good news is that just about anything your food comes home in from the store can make a great planter with a few drill holes. I once got a whole big bowl full of strawberries from plants I potted in some plastic Folgers coffee containers, and my habaneros went berserk (in a good way) when they were planted in the plastic bucket my dog's treats came in. It wasn't attractive looking, but it worked like a charm! If you don't like the look of your homemade planter, you can always put it inside a cute, decorative pot that doesn't have drain holes to catch any leaked water, too.

POTS: Pots come in a wide variety of shapes and sizes. While I always recommend people erring on the smaller side for their indoor plants to avoid root rot, I typically suggest the opposite for people who are gardening outdoors with them. This is because, while in the summer sun, the root systems will grow far and fast for some of these plants, so it's great to give them some room to spread out.

That said, some plants, like pepper plants, grapes, and tomatoes, all love warm roots and do best when their root structures stay above a certain temp. This makes a raised bed or a container preferable in many cases to in-ground beds.

Plant pots are made out of all types of materials, including plastic, metal, porcelain, and clay.

Terracotta pots are the porous orange/brown clay pots you most commonly see associated with planting on television and in magazines. While these pots are somewhat fragile, they do have a very interesting property that most plastic and porcelain containers don't: their clay material can wick away excess moisture and allow more oxygenation for your plant's roots. They also don't leech microplastics into your food the way plastic pots do.

49

All things considered, plant pots are a fantastic way to contain your fruits, vegetables, and pollinator flowers in an attractive way.

FIVE-GALLON BUCKETS: The five-gallon buckets you can buy at any major hardware store make a great planter with a handful of holes drilled in the bottom for drainage. They allow for deep roots and work especially well for tomato plants, small carrot patches, and fig trees.

These are cheap and effective. My only issue with them is that they are made out of plastic. However, there are food-grade versions of these exact same buckets available in an array of colors from companies like ULINE.

KIDDIE POOLS: I once knew a woman who had one of those small kiddie pools for her dog to splash around in in the summer. When the dog passed, instead of throwing it away, she poked a couple of knife holes in the bottom, filled it with dirt, and planted strawberry roots. Since strawberries are thirsty plants, the dialed back drainage kept the plants nice and moist, and she had enough strawberries by the end of her first season to make a quart of jam. It just goes to show, nearly anything can be a planter.

WINDOW PLANTERS: Window planters are typically long, rectangular planters that are intended for use on a windowsill or to be connected to the bottom edge of a window. They are most often made out of plastic, but I have seen some made out of metal and wire with a coconut coir insert. Window planters, also called window boxes, are some of my absolute favorite planters in existence because of their versatility and low profile. You'd be astounded by the things you can grow in a window planter box.

Image 20 - Window planters zip-tied to the top and middle railing of my rental house. This kept them off the landlord's grass and in the full sunlight.

Window planters can be fastened to just about any surface imaginable. They can also line existing raised beds. I even fill some of mine with pollinator flowers and move them around in my garden to draw the bees and other beneficial bugs right to my flowering plants. It works great. I can't recommend these highly enough.

GROW BOXES: Grow boxes are great for any type of garden. They are also a truly fantastic option for renters. A grow box is a self-watering container, often square or rectangular. They are available at just about any major online retailer. These are food-grade plastic bins, often with rolling casters, that are especially great for balconies and patios because they don't have holes in the bottom. They don't leak everywhere or rot the wood beneath by leaving it damp and soaked all the time. The water also doesn't dribble off your balcony onto whoever or whatever is below.

51

Instead, the excess water runs off into a reservoir at the bottom of the planter, wicking moisture through the overflow holes so that the plant doesn't drown. The plants rest above the reservoir screen so that the roots can reach down, sense the moisture, and drink from the well like a straw. The screen also provides great oxygenation for the roots as well, something that can be troublesome in in-ground garden whose soil is too tightly compacted. This oxygenation encourages faster plant growth, too, which can be a

lifesaver in northern areas where growing seasons are often a fraction of the length of somewhere closer to the equator.

The neat thing about these grow boxes is that they are a great way to patio garden in a way that still encourages a deep root structure, which many plants need for stability and nutrient absorption. You also have the freedom to move them around and rearrange them as needed.

Another benefit is that, if you move to another home mid-summer, you can take your plants with you without losing your garden or disturbing the roots.

In these, you can grow anything from massive cherry tomato vines to squash to watermelon. You can even grow root crops like beets, carrots, and potatoes in them. Plus, plants like peppers that like warm roots go absolutely nuts in these.

Image 21 - A photo of my father's grow boxes lined up side by side. Each one is chock-full of peppers.

TIERED A-FRAMES: These A-shaped tiered frames with long, rectangular boxes are a fun way to do some suspended container gardening. While there are many iterations of this design, the concept is the same. These are particularly nice for anyone with a back or knee injury or anyone who is trying to be extra cautious of burning their grass in order to get a security deposit back. Crops that do exceptionally well

in these, in my experience, are lettuce, spinach, strawberries, garlic, peppers, basil, cilantro, and pollinator flowers.

53

VERTICAL WALL PLANTER: I love the concept of these, but have had zero luck with them. Nevertheless, I want to put them on your radar in case they are a good fit for you. Vertical wall planters are long sheets of grommeted felt full of pockets that can be strung up or mounted on just about any fence or flat surface. Each of the pockets gets filled with dirt and seeds. Then, it must be watered frequently.

The real upside to these is that, when they work, they take up no ground space, and everything stays up out of the way, so nothing can get trampled on. It also doesn't burn your grass, which is another pro if you are a renter.

The downsides are that the felt pockets are so porous that they have the be watered constantly to keep from drying out. The pockets are often not deep enough for anything other than herbs, and the water rarely penetrates the soil. I would mostly recommend these for succulents.

HANGING PLANTERS & BASKETS: Hanging planters are fabulous because they take up *zero* space on the ground, which is typically what we often think of when we think of a "garden." In the store, they typically only contain annual flowers, but a wide variety of fruits and veggies can thrive in a hanging planter.

Some hanging baskets are plastic or porcelain, which hold more water.

Other common hanging planters are made of wire and lined with a coco coir soil holder. Those are the most common types available at the big box stores, but, much like the felt wall planter, they leak like a sieve. They are great for plants that either don't like a lot of moisture or for gardeners who like to over-water or even water every day.

Things that I've had do very well in a hanging planter include: strawberries, tomatoes, blueberries, & hot peppers.

SHEPHERD'S HOOKS: While technically not a container, the shepherd's hook is a great accessory for any garden, large or small.

A shepherd's hook is a metal bar that has a couple of spikes on one end and a curved end on the other, much like the staff a sheepherder might carry.

Once a place has been chosen anywhere in your yard, press the spiked bottom into the soil with your foot and hang a plant from the curved end. These hooks are a fabulous addition to any yard or garden because they can be moved at any time.

Some great uses for them are to hang a basket of flowers beloved by pollinators above or near your flowering veggies. This draws the bees and butterflies to the flowers, and while the insects are there, they have a higher likelihood of discovering your blossoming fruit and veggie flowers, too! This brings not only beauty to that section of your garden, but also results in higher yields for your harvests.

Alternatively, you can fill the basket with basil, rosemary, marigolds, or nasturtiums to ward away pests or provide partial shade for a plant that is getting too much sun.

FELT GROW BAGS: Felt grow bags are a great way to grow crops, especially ones that like "dry legs" as they tend to leech out excess moisture pretty well. These are not recommended for thirsty plants (for a list of thirsty plants to avoid planting in these, see the chapter called *The Flood* later in this book)

That said, felt grow bags can be a stellar addition to any garden, especially temporary ones, such as a rental property.

I find that these are perfect for carrots and the companion planting duo of potatoes and lettuce. The lettuce shades the soil, and the potatoes have room to grow and expand. Just make sure you leave enough room to heap some more dirt on once you harvest the lettuce so that the potatoes stay under the soil where they belong.

NO-BEND RAISED BEDS: I call these no-bend raised beds because you really don't have to bend over to work in them like you would in one of the aluminum raised beds shown earlier in this chapter.

These can be made out of wood, metal, or plastic. This style of raised bed is perfect for gardeners with bad backs or knees, those who have suffered recent injury or inflammation, or for someone who just desires their tool storage and plants to all be in one place.

You can grow a *lot* in these no-bend raised beds. Just make sure that the one you select has good drainage, or drill some extra holes in the bottom if not. You don't want those beautiful plants to get soggy and waterlogged. Air is important for healthy roots.

SHOES: I know, I know. What in the world? Shoes? This was just a fun one I thought I'd add because I tried it a long time ago and it actually worked. One of my gardening books said you could grow a watermelon in an old shoe as a way to illustrate that anything could essentially be turned into a planter. Being someone who loves to experiment like a mad scientist full of curiosity, I thought I'd give it a try. I had an old pair of sneakers I was about to throw away, so I packed them with dirt, set them on the side platform of a table near my garden, and jammed a couple of seeds in each.

Much to my surprise, a few weeks later, I had a four-foot watermelon vine growing out of each one. It seems that the cushioning inside acts as a sponge, holding some of the excess water that the roots can slurp down when needed.

Full disclosure: I went on vacation, and no one thought to "water my shoes" in the yard, and they were D.O.A. when I got back. However, it did grow some mighty fine vines. While I wouldn't push for this method because some people thought it was weird and hypothesized that any melons that would have grown from it would have tasted like a wet sock, the point I'm trying to make is that anything can really be a container.

CREATIVE PLANTERS

You're probably laughing at the heading of this section. A pair of old shoes wasn't creative enough? If you think a pair of *Reeboks* is weird, all I can say to that is:

There's more where that came from!

I will be brief in this section. These are not things I recommend for every gardener, just the ones on a super-tight budget. (Just keep in mind that many of these methods will include the nanoplastics I harped on earlier in this chapter, so use them only if you decide it's right for you and your small-space garden.)

STRAWBERRY & CHIVE HAMPERS: I've personally used this method and it worked well. I only stopped because my yard was starting to look a little *Beverly Hillbillies* with all the mix-and-match planters I had all around. Once I moved to a home where I could have an in-ground garden, I gave up many of these methods for the sake of streamlining everything.

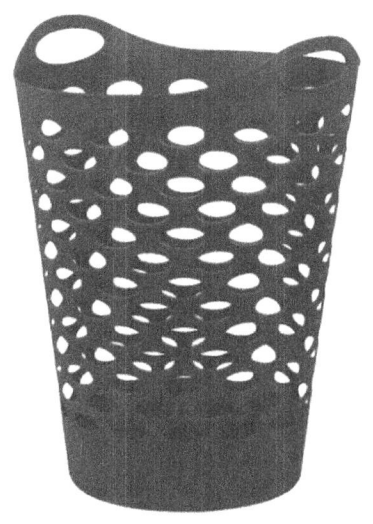

This is a cool method where you buy one of those cheap laundry hampers with all of the wide holes. Then, you take turns lining the outer edge with straw or mulch and filling in the center with potting soil. Then, you insert your strawberry crowns through the holes and alternate in some chive seeds or onion bulbs on the top and through the remaining holes. Water vigorously.

When the strawberries start to grow leaves, you will barely see the hamper beneath, depending on how many strawberry crowns you added.

Then the chives and/or onions, as companion plants to the strawberries, will make the berries sweeter while maximizing your space. As a bonus, you can also harvest chives all through the season.

The main downside to this method is because of the shape, water often heavily coats the top half and never reaches the bottom half. So if you want to try this method, I recommend either getting a small hamper or only planting your crowns around the top half. Because of the funnel shape, too, if you live in a windy area, it might behoove you to fill

59

the bottom with rocks or bricks before you start adding dirt and hay. This should keep it from toppling over. You also might have to give it a spin once a week or so, so that your plants are all taking turns in that southern-facing sunshine. With this method, next thing you know, you'll have strawberries draped all over it, ripe for the picking!

CANVAS GROCERY BAGS: I have never personally tried this method, but I knew a woman once who swore by this method for growing lettuce. This grocery bag method is an easy, disposable way to grow things such as lettuce, spinach, peppers, carrots, potatoes, and small root crops like radishes.

Simply fill it with soil, plant your seeds, and water vigorously, just as you would a felt grow bag.

FOUND ITEMS: I once turned a rolling cooler that I found by the side of the road during Connecticut's bulk trash week into a planter. I took it home, drilled holes in the bottom, and filled it with layers of leaves, dirt, and compost.

My boyfriend chuckled at my somewhat questionable find.

However, that season, I planted blueberries and strawberries, which did phenomenal in my

60

homemade no-bend raised bed. I was even able to keep my spare supplies on the lower shelf to save space.

The next year, I moved all my berries to a dedicated fenced-in berry garden elsewhere in my yard, and I filled the rolling cooler with pollinator flowers. It was a wild burst of color that I could roll around and place wherever I wanted it.

As an extra perk, it came with a built-in bottle-opener that comes in handy when I'm enjoying a drink and watching the sun set over my tiny suburban farm. It just goes to show, anything can be a planter if you want it to be.

GRILLED VEGGIES: I love to upcycle. I hate the idea of things piling up in the landfill if they don't have to. A few years into my gardening journey, my propane grill bit the dust. I had used it for so many years, for so many cookouts, that the bottom literally fell out one day.

Instead of dumping it out on the curb, I stared at it for a couple of weeks off and on, trying to figure out exactly how to turn it into a planter. After all, it had sturdy shelves on the sides for some potted plants and a lovely little closed-off section underneath for all of my spare plastic pots and bags of soil.

After wracking my brain, I bought a three-dollar string trellis and took out the grill plates and removed the lid. I folded the trellis in half and attached it to every hole that I could find on any of the four sides using zip-ties. Then, I laid in about two yards of tulle from the fabric store, which, at the time, cost me about five dollars. (For those unfamiliar with tulle, it is the fine mesh that ballerina tutus are made out of.)

After the tulle was in place, I carefully heaped in the dirt. The string trellis suspended the dirt, and the tulle kept everything from washing out the first time I watered it. It now had a base where water could freely run out, and the dirt was where the grill plates would be.

I planted my seeds, and everything took off like a rocket. The funny thing was, every time I had people come over after that, they always commented about how much they loved that planter and how unique it was. Eventually, it even inspired one of my friends to do the same with hers when it went kaput.

The grill planter worked like a charm, and I got a bowl full of hot peppers out of it that first year.

Image 22 - A snapshot of my unique barbecue pit no-bend planter. My Jack Russell, Cindy, used to always find her way up onto it like a squirrel.

62

HYDROPONICS

Hydroponics is a grow method that requires zero soil. Instead, it uses mineral nutrient solutions that are dissolved in the water for easy uptake by the plant's roots. It uses less water and space than a traditional garden. It also allows for faster growth, great fruit production, and the biggest perk of all is that it can be done year-round, whereas most gardens are only capable of producing for your grow zone's season.

The downside is that it has some somewhat larger startup costs, it isn't maintained like a regular garden, isn't subject to a lot of the same problems as a typical outdoor garden, requires regular monitoring, and, to do it year-round, it requires either counter or floor space indoors plus constant electricity for the LED lights and water pumps. It is also vulnerable during outages without a backup generator.

I could write a whole book on this method (and someday I might), but if it interests you to grow healthy fruits and veggies in water, I suggest doing a deeper dive into this method. I've tried it a little bit. It was certainly fun to dabble in, and there are some really ingenious methods nowadays for growing strawberries and lettuce this way, especially.

GROW UP! GROWING ON ARCHWAYS

Most people, when we are talking, tell me, "Oh, I'd love to have a garden, but I don't have the space." This is often an exaggeration, especially when there are so many great options for growing up and overhead. I have seen gutters converted into long hanging strawberry planters. I have seen hanging planters galore.

Then there are also some of my favorite methods of growing off the ground, which are on archways and overhead trellises. Only certain vining plants will work well for these types of growing, but they do work, and they do add a bit of flair and whimsy to any yard or garden.

Plants that work well for this sort of thing are:

- Grapes
- Watermelons
- Banana melons
- Thornless blackberries
- Cucumbers
- Snap peas
- Climbing green beans
- Cherry tomatoes
- Butternut squash
- Morning glory (*pollinator*)
- & loofah.

Two years ago, I decided to give my grapevines some structure. I bought about twenty dollars' worth of PVC pipe from the hardware store and took a little leftover rabbit wire from my old fence, and in about three hours, I had a frame built and an extremely solid, wired structure for my grapes to climb on.

Then, months later, after the grapes had reached the top, I decided to extend their space overhead by installing two more heavy-duty PVC pipes in the back raised bed, fastening them to my chain link fence, and then using paracord to attach a string trellis from the archway to the back pipes.

Image 23 - My twenty dollar PVC garden archway before I added the rabbit wire or the overhead trellis.

Image 24 - An overhead trellis for my grapes and blackberries made with five dollars worth of trellis and some paracord.

65

I did all of this by myself in a few hours. Again, I am not saying that to be braggadocios. Instead, I am trying to remind you that whatever garden project you put your mind to, *you can do it*!

Image 25 - my whimsical grape archway in full bloom just three months after I built it.

Image 26 - Grapes growing up the side of the archway. They are very easy to harvest this way when they are ready.

Image 27 - The view from underneath the archway. Grapes everywhere!

MOW-FREE HELLSTRIPS

Note: If you rent, you might want to skip this chapter (that is, unless you have a really cool landlord.)

Want to cut down on mowing and attract a boatload of pollinators, all while while adding a blast of color to your neighborhood and providing a safe-haven for beneficial critters? Then, planting in your hellstrip might be just the ticket!

Image 28 - Marigolds and black-eyed Susans in my hellstrip.

68

What is a hellstrip, and why is it called that? These are the narrow strips of grass or vegetation between the sidewalk and the street outside your home. It is also sometimes called a "parking strip" or a "tree lawn," depending on where you live and the local vernacular.

They are traditionally known as "hellstrips" though because they have an environment of generally harsh or "hellish" planting conditions like soil compacted due to foot traffic, reflected heat off the asphalt and concrete, and the shallowness of some. Often, this soil lacks nutrients or picks up excess snow salt splashed onto it in colder areas during the winter.

So why am I talking about this strip of grass if it's such a planters wasteland?

Because I don't believe it is.

In fact, I *know* it isn't firsthand.

Before this year, I didn't know I was allowed to do anything but mow and/or edge my hellstrips because, technically, they are considered *city property*. It wasn't until I chatted up a local plant fanatic highly involved with my city that I found out not only can I use this space for planting, but some cities, like the one I live in, even have *initiatives* and sometimes even *rewards* to encourage homeowners and landlords to do so. This is because, when filled with non-invasive species of plants, they provide homes and food for the local insects and fauna, feed pollinators, provide shade and visual interest, help absorb flood waters, and more. This friend showed me *his* hellstrip two blocks away from my house, and it was loaded with milkweed, blueberries, coneflowers, violets, trees, and all kinds of other great stuff. Strolling down the sidewalk by his house felt like a walk through a lush wildflower meadow, even though we live in a small beach town.

Seeing this encouraged me to get creative with my own hellstrip. After all, I have one of those annoying corded lawnmowers that I have to keep re-plugging into different outlets just to reach the strips on both sides of my corner property. I had a bag of marigold and black-eyed Susan seeds leftover from my garden last year and figured, *Why not?*

I took twelve minutes to till up one of the strips with my electric tiller, scattered my flower seeds, mulched it, and kept it watered for about two weeks. No fertilizer. No compost. No frills. Just till, scatter, mulch, and spray. Then, I don't think I ever watered it again for the rest of the summer, and that thing bloomed with ferocity. Two months post-planting, I was getting just as many compliments on my hellstrip as I was on my lush vegetable garden!

Having all of these free flowers blooming attracted more bees and butterflies and gave me the added bonuses of warding away some of the critters that don't like the smell of marigolds (like the family of skunks that lives under my neighbor's porch. They didn't tunnel into my garden *once* this year because of those marigolds!)

Planting in your hellstrip is a great way to visually liven up your property, attract pollinators that will, in turn, boost your garden's yield, and even increase your garden space.

It doesn't have to be flowers, either. Plant some gorgeous cabbage out there and watch locals ask, "What is that gorgeous plant?" Then, at the end of the season, harvest it and eat it, donate it to a local soup kitchen, or turn it into compost to feed your garden again next year!

Some considerations when planting in your hellstrip:

- Don't get too attached. Because it is the city's property, if they decide they want to widen your road and cut into that area, they can. This *is* a very rare occurrence, though.

- Check with your city's local ordinances, often listed on your city's website. Some places have rules about what you can plant, or they also may have requirements about the maximum height of the vegetation in the hellstrip that are handy to know upfront.

- Ask your city for something to plant! Many times, the city itself will donate a tree or native plant to put in your hellstrip.

- Plant the right type of plants. Preferably, nothing invasive to your area, which can easily be found with a quick internet search about your town. By installing drought-tolerant native plants, you will ensure that you rarely have to water anything once they are established. That said, if you want to plant thirstier plants like strawberries and such, know that you will have to water them more and wash them very well, as they may have come into contact with animals and shoe soles at some point.

- Feel free to get creative with things that are ornamental or edible. This coming growing season, because the soil here in southern Connecticut is perfect for zucchini, I plan to plant a patch of it in my front yard hellstrip with a sign permitting passersby to harvest and eat any that they see and want for free. All it costs me is a couple of seeds, a little mulch, and a few seconds of water each day while it's getting established.

71

CONTAINER ORCHARD

Maybe you're still on the fence about an in-ground garden or raised bed, but you still want some delicious fruit that you grew. No matter your situation (even if you rent or garden on just a patio), you can have your very own container orchard!

While, as a general rule of thumb, fruit trees *usually* do better when planted in the ground, some fruit trees still do very well in containers. Some even *thrive!* Here is a list of some fruit trees that do well in containers, such as food-grade five-gallon buckets.

FIGS: I'm starting with the best one first. Not only do these plants do well in containers, but some actually do *better* in a container. There are over a *thousand* varieties of fig trees in existence, most of which are self-pollinating.

Depending on where you are and what your grow zone is, growing these delicious fruits in a container may be the *only* way to grow your own, as many varieties of fig cannot handle cold temperatures. Some northerners in colder climates tend to bring these inside in the fall or overwinter them in a garage wrapped in fleece, burlap, or other insulation once the temperatures start to dip. I personally used to *only* grow these in containers. They are my boyfriend's favorite fruit, so for years, we always brought the pots back inside in the fall and stuck them back in the yard come May.

POMEGRANATES: The pomegranate is a fruit in the berry family, native to the Middle East. These are often hard to find, so if you love these as much as I do, it makes growing them such a treat. Imagine being able to have your own fresh supply close at hand!

It is important to note that pomegranates do require a fairly large pot once the tree starts to mature, but these delicious red delicacies are a great addition to any container orchard, and there are some dwarf varieties available that take up surprisingly little space.

AVOCADO: This is one of my absolute favorites to grow. Unless you are in a really warm or tropical climate year-round these will have to come inside in the winter. They are easy to care for with some well-draining chunky soil (more on that later) and a little bit of granular citrus fruit fertilizer. The thing I love about avocado trees is the foliage they produce. In my opinion, these visually beat a ficus or palm in the beauty department hands-down.

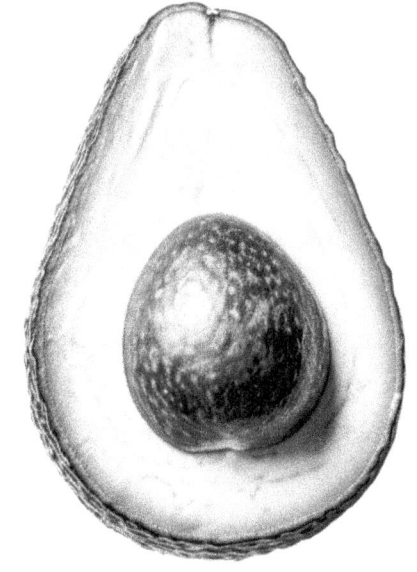

You can grow them from a pit from one of the avocados you buy in the store, but if you are expecting them to produce lots of that yummy green goodness you get at the supermarket, you might be a *little* let down. If grown from an avocado pit, it can take one of these trees around five years to produce its first avocado, *if* it ever produces one at all. If you want the deliciousness *and* the beauty, I recommend going with a grafted avocado tree that you can buy online from a reputable retailer. When these arrive grafted, it shaves years off that wait time, and the plants are often strong enough to hold the weight of the fruit it is bearing, whereas some of the homegrown ones can snap or topple if you haven't shaken them frequently to simulate vigorous winds or topped them a few times when they were small to encourage a fatter trunk.

I have purchased grafted avocados from online stores before, but I honestly do love watching them grow from a pit. They are gorgeous trees. I don't stick toothpicks in mine and keep them half-submerged in water. I just stick the pit in a pot of moist dirt, cover it with a thin layer of mulch, and stick it in a sunny spot. In a few weeks, I always see beautiful growth.

MANDARINS, SATSUMAS, LEMONS, AND LIMES: These citrus fruits are

all fabulous choices for your container orchard. They all do quite well in a pot, and there are even some varieties that are just stunning, like the variegated pink lemon tree that puts out... You guessed it: Lemons that are pink instead of yellow inside. These are perfect for making fresh-squeezed pink lemonade.

The upsides of all of these are edible fruits that you

74

can enjoy year after year. Another is that the blooms smell incredible. I adore the way my lemon tree smells when it blossoms indoors. It smells better than an essential oil!

There are two downsides to growing citrus indoors. The first is that, since 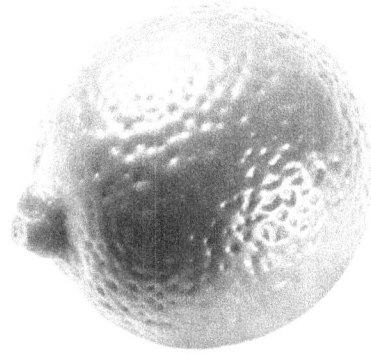 citrus fruits are all tropical plants (this being the reason why Florida is famous for its oranges) is that if you do not live in a tropical paradise and you have cold winters like I do, these all have to come inside. I also don't recommend overwintering them in a cold garage like you can with a fig tree. I recommend bringing them into the same living space you spend the cold months in, instead. They need to stay warm *and* humid and far from drafty windows and blazing radiators while still getting light. The best place, if you have room, is in the bathroom with a little LED light overhead. Warm showers and baths keep the humidity high and the plant happy.

The second downside to growing citrus trees inside is that lemons and limes have a built-in mechanism for keeping birds from picking off all of their young fruit: they grow long, super-sharp spike-like thorns all over. These aren't short, fat hooks like those of a rosebush. These can be around an inch long, thin, and super rigid like a sewing needle. If your house is cramped and you have to brush up against the plant in the winter, you could end up getting poked. The ways around this are

75

to cut the spikes off with scissors as the plant grows, or to keep it in a spot where you won't come into contact with anything but the pot.

COFFEE PLANT: These are beautiful plants with forever-glossy foliage. These are a stunning addition to any home orchard, though I mostly recommend these for their visual value. Many conditions have to be perfect for them to start fruiting. If you are lucky enough to get one to fruit, rejoice. Inside the fruit lies your very own home-grown coffee bean!

Personally, I love growing these. They are often available in the little five-dollar area of plants in tiny pots at your local big box hardware store. If you size them up to a bigger pot each year, these will turn into gorgeous trees before your eyes. If you keep them in small pots, they will stay contained and not outgrow your living space. Ease them out into the sun in the late spring and watch their growth explode.

PEACHES: Only certain varieties of peach trees can stay in a pot, and eventually, years down the road, most will need a pretty large pot (by that, I mean twenty-five gallons or so.)

If you want to grow your own peaches at home, you want to look for ones from online merchants (Like the Arbor Day Foundation) that are either dwarf or patio varieties.

These include:

- Bonanza
- Bonfire Patio
- El Dorado
- Honey Babe
- & others

DWARF CAVENDISH BANANA:

This is one of my absolute favorite indoor/outdoor potted fruit trees. If you live in a warm or tropical climate, you can keep these outside and they'll flourish. However, if you're like me and it gets cold in the winter, it will need to come inside.

You can start a small one of these in a five-gallon bucket, and it will do well for a year or two. Then, it will be time to size up to a bigger pot.

These plants can get large, so they can be tricky in homes or apartments with really low ceilings or very little floor space, once you've sized up to a much bigger pot.

I have known people who have put these near a window, and after a year or two, they suddenly have a bunch of bananas growing!

I personally had one for about three years that I raised from a pup. I chose to sell it when I moved from Louisiana to Connecticut, but I really loved that tree and plan to have another. While I wasn't lucky enough to get it to fruit in that short amount of time I did get loads of compliments on how gorgeous and stately it was in my living room.

Another cool upside to the banana tree is that they produce pups regularly, which simply means that baby banana trees will sprout up through the soil near the base of the original. You could always let it grow in the same pot, but with a bread knife, you can also separate it from the mother plant and put it in a smaller pot. Then, you have a second tree for free to re-gift to a friend, grow in another room, sell or use as compost.

GREENHOUSES

While this is another chapter that I could go into an insane amount of detail on, I'll save the deep dive on greenhouses for another book. This is a guide to teach you how to grow a lot of edible stuff in a small space. If you have a greenhouse at your disposal, you likely aren't gardening in cramped conditions. However, I still wanted to include a little about them because I think understanding why they are the preferred method of a lot of growers will teach you a few things about your container, raised bed, and in-ground plants, too.

So what is a greenhouse?

A greenhouse is a structure, usually made from a wood frame with lots of glass, clear PVC, or plexiglass, made for growing year-round. Sometimes these are also made with steel or aluminum frames or with corrugated plastic or fiberglass sides and roofing. They range from the size of a suitcase to the size of a warehouse. The concept is pretty much the same no matter the size, though.

79

Greenhouses typically have a door for access as well as ventilation slots, mesh screens, windows, and/or flaps that open to keep plants from baking. Depending on the size of the greenhouse, there is sometimes electricity and water running to it as well.

Through the many transparent windows, the greenhouse traps the sun's heat to create an environment with sufficient warmth. This allows for greatly extended grow seasons, bright light, year-round crop harvesting, and protection from extreme cold and inclement weather. This kind of growing environment supports delicate or exotic plants and tropicals, along with your average fruits and veggies. The structure creates a liveable environment for crops that would normally not survive outdoors. They are especially popular in northern growing zones where the gardener's season is cut short by cold weather, thereby extending the grow window.

Image 29 - The interior of my father's greenhouse.

For the small, portable versions, these can be a fabulous place to start your seedlings indoors that cost only a few bucks.

For the larger ones, however, there can be a variety of associated costs, depending on the size of the structure. These include the cost of labor and materials; electricity to run air conditioners in the summer and heaters in the winter; electricity to run misters, humidifiers, or water features; the cost of store-bought pollinators like ladybugs; and the monthly cost of irrigation.

Image 30 - Another interior shot of the raised beds in my father's greenhouse in Wyoming.

81

DOWN AND DIRTY: GREAT SOIL

What is your type of soil? Do you know? This kind of knowledge can help you get off to a great start with your in-ground garden. Note: If you only plan to container garden with bags of dirt from the hardware store, feel free to skip this chapter for now, as it mostly deals with in-ground gardens and raised beds that will be filled with soil from your yard.

Sandy. Clay. Loamy. These are the three main categories of soil types. Let's briefly go over what each one means.

Soil texture is determined by the size of the particles in it. Clay, silt, and sand are all mineral particles made from rocks broken down over thousands of years by environmental factors like animals, rain, and wind.

Sandy soil drains quickly. It has large particles and is often low in nutrients.

Clay soil compacts easily, retains too much water, and has particles so fine, you often can't see them individually with the naked eye.

82

Loam is _ideal_ for gardening. It is a mixture of clay, silt, and sand that retains some moisture yet still drains well. It is often rich in nutrients.

What type of soil do you have? There are a few easy ways to tell.

One is the Squeeze Test. This is where you till up a little dirt from the spot where you plan to garden, moisten it a little, and squeeze it in your hand. When you open your hand, look at the soil.

If it stays in a clump like dough and is dense to the touch, you have a soil that has a high clay volume. If it crumbles and falls apart like at the beach, your soil has a high sand volume. If it holds its shape but crumbles when you give it a gentle poke, you have loamy soil.

There is also the Mudshake Test.
◆ Fill a container with straight sides (like a mason jar) two-thirds full of water.
◆ Add a small amount of laundry detergent.
◆ Add enough dirt from your garden to fill the jar.
◆ Close the lid and shake it like a cocktail.
◆ Once it has been vigorously shaken, put it somewhere it won't be disturbed for a few days.

Once you check back on it a few days later, you'll notice that the particles have separated and settled into clear layers. The sand drops to the bottom. The silt settles above that. The clay settles above that. (Note: organic matter will float and should be disregarded for the purposes of the test.)

Now you can tell roughly what percentage your soil is of each material just by looking at the jar's results!

SOIL METERS

Soil meters are a cheap way to learn a lot about your soil. These are fairly inexpensive and readily available, and I certainly recommend having one around, as I mentioned in the chapter about tools earlier in this book.

Most soil meters measure pH, light levels, and moisture content through dual probes. Simply jam the probes into your soil and select what you want to meter.

If you'd like to know more about your soil than what a soil meter and a Squeeze Test can tell you, professional testing services are also widely available and are similar to the testing services for aquarium water and swimming pools.

With professional testing, you can have the pH tested as well as the levels of magnesium, calcium, potassium, phosphorous, boron, manganese, and zinc.

Personally, I have never used a professional testing service. However, if you feel like you're doing everything right and your garden is still struggling, this might be a good option for you.

SOIL PH

The pH of your soil is actually kind of important. This relates to the acidity or alkalinity of the dirt that you are growing in. A pH test measures the ratio of hydroxyl versus hydrogen ions that are present. A neutral pH is considered pH 7. When the hydrogen is higher, the soil becomes acidic, usually ranging from pH 1 to pH 6.5. Conversely, when the hydroxyl is higher, the pH is considered alkaline, ranging from pH 5.8 to pH 14.

The majority of vital nutrients that your plant needs become most soluble between pH 6.5 and 6.9. This is why the majority of plants really do well when your soil is in this range.

If your pH is off, try planting fruits and vegetables that prefer alkaline or acidic (whichever yours is) or amend the soil (More on that coming up.)

If you have chosen to amend your soil, it is important to remember to do so *slowly*. If you try to do too much too fast, you can actually tip the scales way out of whack with the over-correction, and it will be just as difficult to grow there. Slow is key. (And by slow, I mean over the course of a *growing season*, not over the course of *two days*.)

ACIDIC SOIL

If the pH of your soil is less than 6.5, you might want to try planting some things that excel naturally in this type of dirt.

If you would rather try to amend the dirt so that it swings back toward neutral, you can add powdered limestone and/or wood ash in the autumn or winter. Again, I advise doing so sparingly so as not to over-correct too much.

Fun fact: the states of the *eastern half* of the United States are more likely to have soils that err on the acidic side of the scale.

Here are plants that do well in slightly acidic soil:

- Blueberries
- Tomatoes
- Cranberries
- Gardenias
- Hydrangeas
- Azaleas

ALKALINE SOIL

If the pH of your soil is higher than 6.8, you might want to try planting some things that excel naturally in this type of dirt.

If you would rather try to amend the dirt so that it swings back toward neutral, you can add pine needles, granular soil acidifier, gypsum, or elemental sulfur in the autumn or winter. Again, I advise doing so sparingly.

Opposite acidic soils, the states of the western half of the United States are more likely to have soils that err on the alkaline side of the scale.

Here are plants that do well in slightly alkaline soil:

- Beets
- Asparagus
- Cabbage
- Broccoli
- Lettuce
- Spinach

- Onions
- Pumpkin
- Squash
- Zinnias
- Lavender
- California poppy
- Cosmos
- Bleeding heart
- and Coneflower

IMPROVING SOIL STRUCTURE

Improving poor soil has its rewards, namely that your fruits, vegetables, and pollinator flowers will be healthier and more productive.

To improve sandy soil: Work a few inches of organic matter like finished compost or manure. You can also grow cover crops or green manures. A green manure is simply a crop that is grown and then tilled right back into the soil before it matures or goes to seed. Great examples of these are vetch and clover, which fix nitrogen from the air. Some people also use rye.

To improve clay soil: These are often nutrient-rich soils, but because of the compaction by foot traffic and the elements, they get packed in too tightly for the plant's roots to access them. You have to break it up. You can do this by tilling in a few inches of organic matter the first year (in the fall, preferably) and then another inch or two every year after that to keep it from re-compacting. Using a raised bed with clay soil is especially good because this method improves drainage *and* keeps everyone's feet from compacting the dirt down over the season.

EARTHWORM CASTINGS

Earthworm castings are often sold by the bag as a soil amendment, and this is another little trick I picked up that completely changed the game for me as a gardener.

Worm castings are essentially code for worm poo. Disgusting, right? Trust me, it actually isn't as gross as it sounds. They just look like soft, black granules. They are an excellent fertilizer that won't burn your plants' roots, unlike some products. These castings offer a *myriad* of benefits. They improve soil structure, slow-release essential nutrients without burning the plant roots, retain water, offer disease and pest resistance, rocket plant growth, and so much more. They are chock-full of nutrients and beneficial microbes that improve your garden's overall health.

Can you garden without them? *Absolutely*. Are they a great amendment? *Yes*. Full stop. Can you get them without having to buy them? You sure can, through either vermicomposting or adding a healthy dose of shredded paper or tape-and-ink-free cardboard to your in-ground garden or raised bed beneath the top five to ten inches of dirt. The worms will come running (well, I guess *slithering*) to the area to feast on the paper, and they will leave you um... *presents*... while they get to work aerating your soil.

Your garden will thank you for it by offering *quite* a harvest.

COMPOST & AMENDMENTS

First, let's talk about compost, or as I like to call it, "black gold." Compost is organic matter (that would normally be considered waste) that billions of microorganisms will work to break down over weeks or months into a soft, nutrient-rich amendment for your soil. This process takes place successfully when combined with a certain level of both heat and moisture. Basically, it is turning your yard leaves, grass, and kitchen scraps into fuel for your plants.

Compost is one of the best things that you can give your garden, and it is absolutely free.

In this chapter, we are going to discuss a few methods for composting, including some that are absolutely free to do with the things you might find around your house or yard, or in somebody's trash. By adding this nutrient-rich compost to your potted plants, in-ground garden, or raised beds, you are using your kitchen scraps and discarded cardboard to not only feed your plants but the earthworms as well. Many of you already know that earthworms are an underrated treat for any garden or raised bed. Not only do they aerate the soil with their squirming and tunneling, but when they expel their waste, they expel their earthworm castings.

Compost is great because it's all stuff that you would have naturally thrown out into the garbage, so not only are you reducing the stress on the local dump, but you are giving your garden free fertilizer back that will, in turn, feed you again.

One of the downsides to composting is that you will end up with renegade plants, more frequently called volunteers, because a lot of times the seeds do not break down during the composting if the temperatures do not get hot enough. When you put a bunch of kitchen scraps together in the heat, the warmth breaks them down and reduces the smell. Many people use a variety of ways to compost, including store-bought composters, which are fantastic and easy to assemble.

However, if you are reading this book, it is most likely because you have very little space to garden, so there are other ways to achieve the same goal with a smaller backyard footprint, should you so desire. One of those ways is to dig a trench and regularly put your food scraps in and cover them with dirt until the trench is filled. Another way is to get yourself a little plastic storage tub or a plastic trash can with a lid and drill some holes for ventilation toward the bottom and a few holes in the bottom for drainage. Put all your food scraps inside and make sure you secure the lid and put something heavy on it, like a wooden board, and then a planter. This keeps raccoons, possums, and squirrels out of the compost bin while also giving you a functional plant stand.

Here are a few tips and tricks that you can do with your compost pile to maximize it or jump-start the process. One great tip is laying bubble wrap over the top and sides of your pile and laying bricks on the edge so that it doesn't fly away. This keeps the compost pile from soaking up all the rain and moisture, and it allows the sun to heat it through the clear bubbles when the sun is out. The bubble wrap also acts as an insulator, locking the sun's heat in and speeding up the process.

Another fun trick is to pour sugary soda in with your compost. I also use spare, skunky beer. Pouring these onto your compost pile helps jump-start any microorganisms that already exist either in the soil or the organic matter itself. Sugary soda increases the acidity as well, and the sugar can actually feed all of the microorganisms inside.

If you don't have a compost tumbler but you have the room for a dedicated compost area, why not use an old tire? If you have such a thing lying around, you can fill the base with shredded paper or cardboard and pile in all of your kitchen scraps and old shredded newspapers. If you put this directly on your grass or bare dirt, the worms will find it and work their way up through the paper and/or cardboard to start eating the compost and turning it into earthworm castings. The sun heats up the rubber and can speed up the process of breaking down the matter inside. Getting your compost out is as easy as tipping the tire onto its side and shoveling out all that black gold.

Old tea, when thrown into compost, can help speed up decomposition and attract bacteria that produce acid, creating a compost your plants will love.

Save those old coffee grounds. Do not throw them out! Throw them right into the compost heap. Any liquid coffee can help decompose the matter, and the grounds create such a nice, rich compost. Alternatively, if you have hydrangeas or blueberries or any other extremely acid-loving plant, you can bury the grounds in a small hole near the root system, or, if you're lazy like me, just dump them right onto the dirt around the plant. The rain and elements will help work it into the soil.

Having trouble with dogs, raccoons, or opossums getting into your compost? Sprinkle black pepper, cayenne pepper, or liquid Tabasco right onto the compost. The smell will ward away animals and make them think twice before scavenging through the pile.

Many things that most people throw away can be used as compost. They will decompose fairly quickly, and you will be able to amend your soil and fertilize your plants with a free fertilizer that takes some of the strain off the local dump. Some of the things that can be added to your compost pile are fallen leaves, scraps from the kitchen like uncooked vegetables and fruits, chicken poo, eggshells, banana peels, spent tea bags, paper towels, and grass clippings. I even put small sticks in mine as well. It all breaks down with heat and time.

If you want those free earthworm castings, these squirmy creatures absolutely *love* cardboard and paper. As long as it doesn't contain a lot of colored ink or glossy photos, paper and cardboard will provide a feast for the worms in your garden and leave castings that are nutrient-rich. (My boyfriend thought that I was crazy the first year that we were together because he watched me shred old bills and junk mail and flatten a bunch of cardboard boxes. I lined one of my new raised beds with it and threw dirt on top. He said it looked like a strange thing to do. It wasn't until the next summer, when we were digging a hole to plant a blueberry bush, that he saw *why* I had done it. The dirt there was loamy, aerated, and rich with worm castings. Digging the hole for the blueberries, we saw out some of the fattest earthworms either of us had ever seen.

I digress, back to the process itself. Composting works best when the pile is kept moist and is occasionally turned with a pitchfork or a shovel, or by hand. How long does it take to compost something? Some people say it only takes 3 to 6 months for the matter to decompose thanks to the microorganisms inside. However, I typically wait one year (or one growing season). But many people leave theirs for two years (especially if it contains quite a bit of chicken feces, because they can burn the roots of small plants if not decomposed enough) because sometimes it takes a little bit longer to break down some of the matter.

Proper compost is soft and almost black, much like that of the forest floor just beneath the top layer of leaves. It feels and looks like moist potting soil when the process is finished. Thankfully, you have leeway and options for building your compost pile.

The main ingredients for good compost are carbon-rich materials like leaves and straw. Then, you also want to add nitrogen-rich items like kitchen scraps and grass clippings. Then, add heat and moisture, and you've got yourself free compost. It really is that easy.

Many books and podcasters will harp over the ratios and percentages of each for good compost, but I am here to tell you, it really doesn't need to be so *scientific*. It's honestly one of the easiest things you can do as a gardener. It's a low-stress, low-effort way to make your next season's garden one that will have the neighbors in awe of your skills.

THINGS TO <u>ALWAYS</u> COMPOST

- Banana peels
- Rotten bananas
- Coffee grounds
- Eggshells
- Fruit and vegetable scraps
- Grass clippings
- Dried, fallen leaves
- Shredded paper
- Cardboard

You might be asking yourself, "Is there anything I can't add to this compost pile?" The short answer is *yes*. There are a few things I do not recommend adding to that compost bin.

Image 31 - My spinning composter. These are amazing.

94

THINGS TO AVOID COMPOSTING

- Barbecue ashes (also called fireplace ash or wood ash) - These are super rich with sulfur oxides, and they can throw off the pH of your garden if applied liberally. Ashes should be used sparingly. I usually grind mine up and sprinkle it through the lawn and in my more alkaline-loving flower beds or near the bases of my trees and bushes, always in small quantities.
- Colorful magazines - Magazines contain ink. That ink can be potentially harmful in high enough quantities if it is being used as compost for edible items that you will be ingesting.
- Dog and cat feces - While manure is actually a good thing for the garden, the feces of certain animals are not recommended, because once in a while, they can contain diseased organisms that are transmissible to humans.
- Citrus peels (i.e., avocado shells or orange peels) - These won't hurt anything, but they take a few years to break down and can clog up your tiller.
- Food grease and cooking oil - Grease can slow the decomposition of your pile and draw unwanted pests.
- Animal bones (i.e., chicken bones, beef ribs, etc) - While bonemeal is great for flowers and vegetables, I don't recommend whole bones if your garden isn't fenced. While they can break down into nutrients that will make your garden thrive, dogs and animals who are attracted to carrion (such as raccoons and opossums) will sometimes try to dig them up and even leave fragments in strange places.
- Tobacco products or cigarette butts - Cigarette tobacco can sometimes contain Tobacco Mosaic Virus. Once it infects your garden, it can do a *lot* of damage. In some cases, to eradicate it, crops cannot be regrown

for five years or so in a space where contamination was rampant. For this reason, when I go to give someone a garden tour of all my lovely veggie babies, I strictly prohibit smoking in or around the garden.

A FEW MORE COMPOSTING TIPS

- To speed up the composting process, put your materials through a shredder, blender, or mower, or finely dice them. It will decompose *much* faster this way. You can also add a few scoops of compost catalyst, such as one of the pre-made granular ones that come in a bag from companies like Jobe's.

- If you turn your pile more often, for example, once a week, you provide optimum oxygen, which fuels the bacteria that break everything down. This will give you finished compost in a fraction of the time versus turning it once a month or less. The spinning composters make this chore something that can be done in literal seconds, versus minutes.

- If your garden seems super thirsty, massage in or top dress with more compost. It can reduce the need to water and retain more moisture for the plants that need it.

HOW OFTEN SHOULD I APPLY COMPOST?

In my opinion, you almost can't have enough. Apply compost to your garden and flower beds as often as possible.

At a minimum, I recommend adding a generous amount at least once every spring. There are plants, though, like melons, strawberries, and tomatoes, that seem to go berserk (in a good way) if you give them more. They'll produce some incredible hauls for you if you keep top or side dressing with compost throughout the grow season.

Half of the nutrients of your compost are made available to the plant the first year. Every year, half of the remaining nutrients are released.

When you consistently enrich your soil with compost every year, your plants end up with a variety of nutrients at their disposal, and you end up with improved yields by letting your raked leaves and food scraps work for you.

NATURAL PEST AND DISEASE CONTROL

Compost is also a natural form of disease and pest control. Well, *indirectly*, that is. It produces fatty acids similar to those in the store-bought insecticidal soap sprays. These acids discourage root-knot nematodes without ever harming your worms or above-ground pollinators.

Since there aren't any nematicides rated for use in home gardens (at least, to my knowledge), compost is one of the *only* ways to actually control these nematodes.

The fatty acids in compost are also toxic to many fungal and bacterial diseases that can destroy your beautiful fruits and veggies, so applying it to your garden is like a win-win-*win*!

VERMICOMPOSTING

Vermicomposting is the conversion of food scraps and organic materials into those castings we talked about earlier by feeding them to earthworms. Red Wigglers seem to be the go-to.

Benefits of this type of composting include lessening waste in your local landfill, improving your soil's health, and creating a nutrient-dense form of gentle, sustainable fertilizer for your crops (all while lessening reliance on pricey chemical fertilizers and reducing the need for pesticides). And, as a cherry on top, vermicomposting processes garbage and waste faster than the more traditional methods of composting.

You can purchase multi-tiered vermicomposters online that are ready to go, but making one is super easy, extremely cheap, and can often be done with things you already have lying around the house!

Image 32 - Above are two variations of vermicomposters available online. The second type, you bury halfway and the worms in your existing soil come and go as they please. If you have the yard space, I recommend those highly because you don't have to buy any worms.

98

HOW TO MAKE YOUR OWN VERMICOMPOSTER

Step one: Create a bin or container with some top and side ventilation and a few small drainage holes at the bottom. I used a plastic storage bin (the kind you store holiday decorations in) and drilled a few holes in the tops and sides. Then, I used a small knife to puncture a few plus signs into the base for liquid drainage.

Step two: Add a five to eleven-inch layer of moist bedding. Add things like shredded paper, cardboard, and/or coconut coir. This is called the worm blanket.

Step three: Add worms. Red Wigglers, if you can find them at your local bait shop, but just about any worm you find in your yard will work, too.

Step four: Feed them. Add in your vegetable and fruit scraps, eggshells, and spent coffee grounds. I recommend avoiding dairy, meat, or grease. To prevent fruit flies, bury your food scraps under the shredded paper. Personally, I just toss my scraps on top because my vermicomposter is outside and I am a little lazy.

Step five: Place the lidded bin in a dark, shaded area. The lid will provide shade and keep the birds from feasting on your wiggly workers. It'll also keep them from baking to death or evacuating through the vent holes because of the intense heat.

Step six: Maintain moisture. The bedding should feel moist like a wrung sponge, not a bowl of cereal or a cracker. If it starts to dry out, sprinkle or spray a little water in. If it is too wet, you might need more drainage holes or slots in the bottom. Dump the bin every few months, pick your worms back out of the finished castings, and put them back in the bucket with some new paper, cardboard, and scraps for another round of feasting. Then, use the finished castings in your garden beds and pots!

For extra credit, suspend your bin and set a catch tray beneath it. Over time, it will fill up with a liquid called worm tea, which is a nutrient-rich bedpan for worms, essentially. It sounds and, frankly, *looks* disgusting, but plants adore this liquid. Dilute it down and put it in your watering can and watch your plants go berserk!

OTHER AMENDMENTS

Perlite is volcanic glass that expands to form a porous, light material when heated. It is white, granular, and feels as weightless in your hand as Styrofoam. Among its other uses in the insulation and beverage industries, it is sold by the bag at most garden supply and big box stores and used in potted plants and gardens to improve root aeration and drainage. It is great if your soil has a high clay volume or if your potting soil keeps water-logging your plants because it keeps the soil loose. Most Perlite found in stores also contains fertilizer that will help keep your hungry plants well-fed.

I particularly love to mix this with soil from my yard to pot my indoor plants, such as my monsteras, because they thrive in chunky growing mediums. Perlite is a wonderful amendment used in seed starting as well, because it allows the budding roots to really get through soil that might otherwise be so compact that it stunts their growth.

Image 33 - A handful of Perlite.

101

Vermiculite is a mineral that creates a flaky, light, granular material when heated. It looks like mica and also weighs next to nothing. This amendment is great for gardening, potted plants, and seed starting because it holds water while simultaneously improving your soil's aeration. It has other uses, too, including fireproofing and insulation.

One last note about vermiculite: Newer brands of pure vermiculite are sterile and safe, but some older ores of it were found to be contaminated with trace amounts of asbestos. If you are unsure if yours qualifies, or just want to be extra safe, you might want to do a little extra digging on the brand you select or take extra precautions when handling it, like gloves or a mask.

Image 34 - A handful of granular vermiculite.

Sphagnum Peet Moss is a soft, fluffy, dark brown material made from partially-decayed moss that accumulates in bogs over thousands of years. This amendment, when mixed with your soil, improves the retention of moisture, increases acidity, and allows for greater aeration for your roots. This dead plant material is highly absorbent.

While I love this stuff, it is vital to remember that this increases the acidity of the soil around it, so make sure you are either using it to amend soil that is slightly too alkaline or that you are planting vegetables and flowers that prefer a higher acid content.

This is a great amendment for potted blueberries, strawberries, tomatoes, and hydrangeas, etc.

Image 35 - A small pile of sphagnum peat moss.

103

FERTILIZER & MULCH

WHAT IS FERTILIZER?

Fertilizer is essentially plant fuel. This is a substance, often purchased in granular or water-soluble powders, that, when added to your soil, provides essential nutrients for plant growth, namely nitrogen, phosphorous, and potassium. Fertilizer can also come in the form of pound-in degradable spikes, foliar (leaf) sprays, and pre-mixed concentrated liquids.

Nitrogen (N) promotes healthy stem and leaf growth.

Phosphorus (P) helps with the conversion of energy as well as root, fruit, and flower production.

Potassium (K) strengthens the plants and makes them more resistant to diseases.

Some great natural household fertilizers include:

- Banana Peel - Contains potassium and phosphorous. Great for home compost, vermicompost, or making a banana water solution to drench plants. Especially great for tomatoes, potatoes, fruit trees, and rose bushes.

- Coffee Grounds - Contains nitrogen and magnesium. Sprinkle a light coating in the dirt or use your spent grounds in the composter or vermicomposter. Especially great for strawberries, cucumbers, potatoes, blueberries, hydrangeas, and azaleas. It also wards away slugs and snails!

- Fish Tank Water & Filter Sludge - This stuff works amazing and, if you have fish, it's free! Your fish tank's water is full of nitrogen and it hydrates as it feeds. Plus, it won't burn your plant's roots. Fish remains, filter sludge, and fish tank water make my garden plants shoot up like a rocket. I swear by using fish waste in my garden. I attribute *much* of my growing success to it, actually! Use it on any plant you want to flourish. (Note: Too much salt can harm plants, so if you have a saltwater tank, I recommend only using the sludge/waste from filter cleanings as opposed to the water itself.)

- Rice water - The water you boil your rice in contains potassium and nitrogen. Once it cools, pour it right into your containers, raised bed, hanging planter, or straight into the ground. Rice water is especially good for pepper plants, cabbage, eggplants, and tomatoes, as well as a wide variety of houseplants, too.

- Tea - Tea contains nitrogen and makes great fuel for your garden. It can also be added to your compost as a "green" compost ingredient. And *chamomile* tea, when used in growing seedlings, can help prevent damping off, which is where the root of your seedling shrivels and dies in its infancy due to soil-borne fungal diseases. So toss those teabags right in the compost or the dirt.

● Epsom Salts - These salts, usually used for bubble baths and physical therapy, contain a high volume of magnesium. You can sprinkle some right into the dirt. Also, if you dissolve some in water and spray your tomato plants and eggplant flowers with it when they start blossoming, the plant will get the message to send out exponentially more and thus giving you far more fruit from the same plant.

WHAT DO THE NUMBERS MEAN?

Once in a while, you will be at the store and pick up a bag of some name-brand fertilizer and see a set of three numbers with dashes between them on the packaging, for example, 10-10-10 or 4-6-3.

That is your N-P-K ratio. These are the bag's ratios of those three primary nutrients we just talked about. They are illustrated in a number that is a percentage by weight. So, in that first example, 10-10-10, that means the ratio is ten parts nitrogen, ten parts phosphorous, and ten parts potassium. That means it is a balanced ratio of all three nutrients.

Sometimes, when I just want to not think about things and fertilize everything in my in-ground garden at the same time, I'll buy a big bulk bag of 10-10-10 and top dress everything in the garden with it. The rest will get washed into the dirt around the plant, and the roots can spread and access it later on if needed.

106

However, another common one that I really love for all of my edible fruits and veggies is a 4-6-3 combo. That's a ratio ideal for fruits and veggies as it prioritizes the nutrients that make for healthy roots and lots of foliage while dialing back a touch on the one that stimulates overall plant health.

Many fertilizers also include other organic ingredients like beneficial microbes that can aid in nutrient absorption.

FISH EMULSION

Yep, I'm going to say it again because it bears repeating (frankly, I would scream this from the mountaintops if I could because it has really been something that changed the game for me as a gardener.)

Fish emulsion is a fertilizer that is made from organic fish byproducts. It looks like runny mud and can be purchased at most major box stores.

The beauty of this stuff is that it is gentle on the plant's roots (unlike many granular or powdered chemical fertilizers that you would find at the store), and plants really go wild over it.

It is mostly nitrogen with low ratios of phosphorous and potassium, usually with a 5-1-1 number grade. Add it to your watering can, pour some right on the soil and water it in, or mix some in a spray bottle and use it as a foliar fertilizer on the leaves.

If you have a fish tank at home, you can make it yourself by using the wastewater collected after cleaning your filters.

WHAT IS MULCH?

Mulch is a layer of material on top of the soil. It is protective, conserves moisture, regulates soil temperature, and can somewhat suppress weeds that can compete for the same nutrients and minerals as the plants. Mulch can either be organic or inorganic.

Organic mulch consists of wood chips, leaves, or tree bark. These improve the soil as they decompose. These will need to be replaced frequently, typically every growing season.

Inorganic mulch consists of rocks, pea gravel, plastic sheeting, or rubber chips. These can look beautiful and, due to the fact that they take significantly longer to break down, will need to be replaced a lot less often. In the case of rocks, probably more like a lifetime. While these look lovely and have their place in landscaping, surely, many of these inorganic mulches can hinder plant productivity, burn or suffocate roots, or break down into microplastics in the soil. I don't usually recommend these for mulching in a garden you plan to eat the harvest of, unless you have done a slightly deeper dive before committing.

In my opinion, the color matters, too. Studies have shown that the color blue in the ROYGBIV light spectrum is crucial for vegetative development and photosynthesis. It stimulates the production of chlorophyll as well as healthy leaf and stem development. It's what keeps your plants, like tomatoes, from becoming spindly, leggy, or too scrawny to support any fruit.

Meanwhile, the spectrum color red helps encourage fruiting and flowering. It is also important for photosynthesis as well as activating the hormones that regulate plant reproductive development.

It is also especially effective when combined with blue light, which is why most of those indoor plant lights you can buy online have three modes: full-spectrum white, red, or blue. If you are growing a monstera and need it to keep putting out those beautiful fenestrated leaves, you mostly need the blue side of the light spectrum. If you are shining it on a lemon tree, that red mode is the one you need.

Note: some of these LED lights simply have a "white mode" which includes the whole ROYGBIV spectrum, or a mode that looks oddly purple, which is the red and blue parts of the spectrum combined.

So why am I telling you about this? You can't control the sun's spectrum. I'm telling you because there is some science that says of the common colors for store-bought mulch (typically black, red, dark brown, or cedar) that red has the added benefit of refracting a little more of that red spectrum back into the air around your plants. I found that when I switched to red mulch, I had larger yields. It's just something to consider when choosing your color.

Image 36 - Storebought wood mulch is a cheap way to deter weeds, keep roots cooler, and lock-in moisture.

109

What <u>SHOULD</u> I mulch?

Mulch helps to hold water and keeps plant roots cooler in the baking sun. Some things will do fantastic if you mulch them. I recommend mulching the following:

- Watermelons
- Cucumbers
- Strawberries
- Marigolds
- Begonias
- Petunias
- Eggplant
- Blueberries
- Blackberries
- Raspberries
- Creeping thyme
- Oregano
- Mushrooms

What do I <u>NOT</u> need to mulch?

Some things don't really need mulch, especially things that grow well in the Mediterranean or Middle East. Here are some things you don't really need to mulch unless you really want to or you find that they are drying out too fast:

- Grapes (this one likes warm, unshaded roots!)
- Figs
- Peppers

- Lavender
- Rosemary
- Thyme
- Milkweed
- Succulents
- Cabbage
- Tomatoes (these like dry legs and warm roots)
- Asparagus

A note about mulching trees and bushes: Please, I beg of you, don't pile mulch up around the bases of your tree trunks or shrubs. I see people doing this all the time (once I even saw a *landscaper* do it, even though he, of all people, should've known better.)

Mounding mulch around your tree's trunk or bush's stems like a "mulch volcano" can actually suffocate the plant's roots, expose it to rot, or make it far more susceptible to disease. This can cause even the largest, most mature tree to suffer or even die.

When mulching a tree or bush, it is wise to always leave a little exposed "moat" of dirt between the mulch and the main trunk or stem so that the plant can breathe.

This moat also creates a nice little well so that when it rains or is watered, the plant has a nice source of water to draw from straight away.

Image 37 - An example of a harmful mulch volcano. Please don't do this to your trees.

111

FENCING

If you are fortunate enough to already have a fenced-in garden area, you can skip this chapter.

What is a fence, exactly? A fence is just a barrier that keeps unwanted feet and a variety of critters off of the crops you worked hard to grow. Can you successfully grow without one? Absolutely. Especially if your fruits and veggies are in hanging planters, you have a great guard dog to chase away the squirrels and skunks, or if you are gardening solely on a balcony, rooftop, or raised patio.

For the rest of us, you will soon learn when you go outside halfway through summer and see your gorgeous tomatoes half-eaten and strewn around your yard like mangled croquet balls or your pumpkins shredded with their remains dangling on your vine... *You need a fence.*

Or maybe it's not a strange varmint at all. Maybe it is your curious dog looking at those dangling concord grapes and licking his chops eagerly.

(Grapes are highly toxic to dogs, by the way. They can cause kidney failure, and Fido doesn't deserve to go out like that). A fence can save time (and lives as well).

Fortunately, there are some extremely inexpensive options for a fence to protect that in-ground garden or raised bed that you've poured your heart and sweat into all spring. There are permanent and temporary options, too, depending on your living situation. So let's explore!

FENCING TYPES

There are many types of fencing, including chain link, brick, wooden privacy, picket, wrought iron, wire, etc. Some are more durable than others. Some put form over functionality. Some are electrified or covered in barbs. Or, if you are in New England, like me, you might even see ones made out of stacked rocks.

Hey, whatever works!

If you own and plan on having a garden for many years, I recommend a chain link fence. They're a good investment. They're common and can be installed in a fraction of a day in some cases. And the big box stores also have the materials to do it yourself if you have the time and inclination. A chain link fence is great because it keeps cats, dogs, people, and deer out (well, unless the deer wants to gallop in. Then, nothing but a tall privacy fence or barbed wire will stop it. The best thing for those is to ward them away. More on that in my section labeled *Varmints*.)

Another pro to a chain link fence is that it lets the light through, so your plants aren't shrouded in the shade of a privacy fence for half of every day. To me, an added bonus is that others can see your garden. This may sound like a con instead, but I assure you that it can be a pro when your gardening inspires your neighbor to do the same in their yard. I cannot tell you how many times a neighbor has seen me sweating away, covered in dirt, bobbing my head to the music in my earbuds, and then, soon after, started a garden of their own and saying that watching me was what inspired them to do so. I swear I've led by example and indirectly gotten at least eight other people to start gardening. It fills me with pride every time. Truly.

If you rent or are looking for a more temporary or inexpensive solution to protect your garden, I *highly* recommend a material called rabbit wire. You buy it by the roll at your local big box or hardware store. At the time of this writing, they are about $50 for a four-foot roll, large enough to fence an entire suburban yard like mine. Then, all you need is a pair of wire cutters or bolt cutters and a few pound-in posts (like the powder-coated steel universal U-post in the image) and a bag of zip ties or wire for fastening everything together.

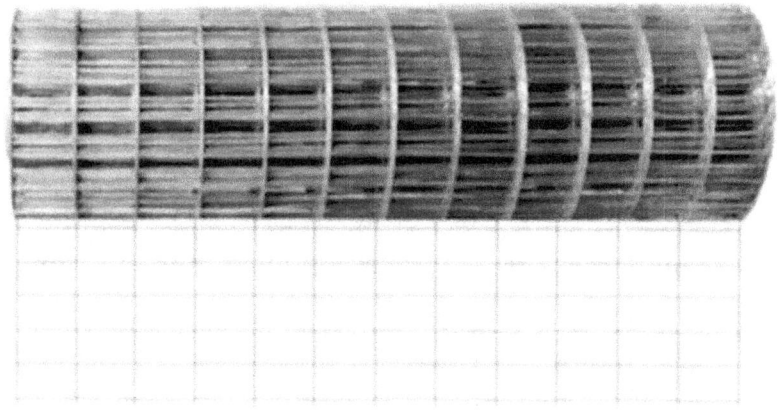

Image 38 - Rabbit wire

114

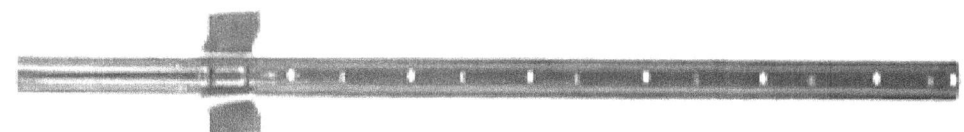

Image 39 - A powder-coated steel universal U-post.

Simply pound a few of the posts into the ground around the corners, cut some rabbit wire to size, and zip tie it to the U-posts. Make yourself a makeshift gate so that you can get in and out, and you've got yourself a protective border.

Rabbit wire is great because it keeps even smaller things out of your garden that might normally get through chain link or picket fences, like rabbits, skunks, large gophers, etc. *This is what I personally use.*

The downside to this type of fence is that some things *still* get through, such as small gophers, small skunks, rats, etc. If those are not very prevalent in your area, this would be a cheap and ideal solution.

If you have a lot of problematic rodents or small creatures, though, you might consider chicken wire instead.

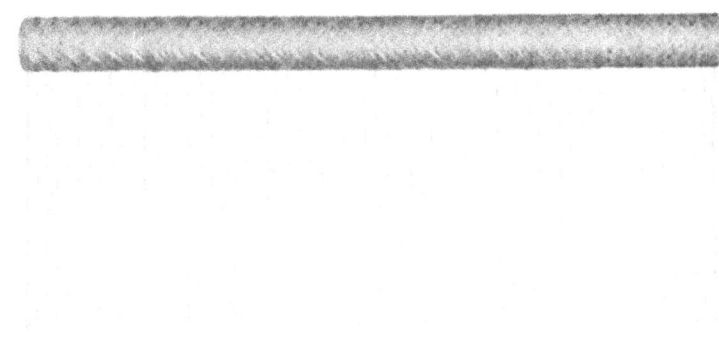

Image 40 - A roll of chicken wire.

115

Chicken wire is a little more expensive per square foot, but it is the same concept as the rabbit wire and will keep out everything but birds, squirrels, deer, rats, and opossums. For those, unless you're willing to build a cage for everything (which I have done for my melon patch because I live in an area overrun by gophers), it is best to ward them away with other methods I'll mention later in this book.

Image 41 - A section of rabbit wire fencing around some of my potted veggies in early spring.

If you're only looking for something to keep out bigger animals and you're on a budget, you can get yourself a few of those $4 U-posts and use a string trellis on its side instead. This netting will mimic a chain link fence and create a border that tells people and dogs, "Hey, you're not welcome in here."

A cage over the top of your raised bed is always an option, too. It's a simple, effective, and inexpensive way to keep animals out of a smaller area.

There's also something called a dead hedge, which I absolutely love. It has its pros and cons, but I think they're great. They're almost medieval-looking. The concept is simple. If you've pruned a bunch of trees and have a lot of little sticks lying around, simply pound some parallel poles, pipes, or branches into the ground every few feet. Then, stuff sticks longways between them. When you've built up enough, you have a totally free fence with some really neat benefits.

So what are these benefits? Well, other than achieving your goal of blocking an easy path for hungry critters to get into your garden, you provide a year-round home for birds and beneficial insects. When the sticks eventually break down, they add compost and nutrients to your soil. They block heavy winds and help give your garden a really unique feel.

Some of the cons are that, while you enjoy the whimsy, your neighbors might deem it an eyesore. While it creates a home for birds, lightning bugs, ladybugs, and bees, it can also create a safe haven for rodents, too. Lastly, I don't recommend putting them very close to your actual home as they could be considered a fire hazard, depending on where you live. That said, I still think they are really unique and natural-looking and worth mentioning.

Image 42 (a and b) - Two dead hedges.

117

No matter what kind of fence you use, you may have to make tweaks or fortify and patch it from time to time, but I assure you that you will keep a lot more of the crops you worked hard for by having one.

Image 43 - A neighborhood doe and her fawn staring at me through one of my old garden fences.

PROTECTING YOUR BORDERS

There is another way to protect your borders that can be used on its own or in conjunction with a fence. I recommend using this method *with* a fence because it strengthens the barrier, offers some additional visual interest, attracts pollinators, and can even repel some of those not-so-beneficial insects.

My recommendation: *a living border* full of trap crops.

What on God's lush earth is *that*?

A living border is when you line your fence, either on the inside or outside, with trap crops or sacrificial plants meant to distract and lure pests away from your cash crops (a fancy way of saying: *the things I'm trying to grow for myself*.) I recommend growing your trap and sacrificial crops on the outside, but either way works.

By planting a row of tight-knit marigolds along the two-inch gap between my fence and the sidewalk, I shored up gaps where the fence doesn't quite go down far enough. When the flowers grow, they block the view of the crops from things scurrying on the ground a bit. They also emit a smell that I personally find pleasant, but many animals like gophers find to be noxious. They look pretty, and they also repel a lot of bugs, too. Lastly, their roots grow around the base of the fence, and it really just tightens everything up.

After two years of being plagued by gophers, the first year I tried a living border out of marigolds, I didn't have a single crop gnawed on by a gopher the entire growing season. It kept the rats and skunks away, too! It drew a plethora of bees and butterflies. And the best part is that I got *constant* compliments about how pretty it looked in full bloom.

Living borders can be made out of a lot of things, but I recommend using things that repel animals rather than attract them. The one exception to this rule is gophers. My father swears by planting a row of what he calls *sacrificial plants* around the borders of his garden. He usually plants cabbage because he doesn't like to eat cabbage (and therefore doesn't care one bit when the animals fill up on it instead of his watermelons and cantaloupes), and it looks gorgeous when it's in full bloom. The idea with sacrificial border crops is that you literally plant them to distract the critters that normally eat your plants by giving them something appropriate to gnaw on instead.

Flowers are always a great living border, but I'd stay away from plants like petunias, which are a favorite among wild rabbits and deer.

To repel animals and strengthen your border, go for marigolds, nasturtiums, lavender, thyme, sage, rosemary, zinnias, black-eyed Susans, or blanket flowers. For sacrificial crops, I recommend cabbage, carrots, kale, or chard.

Image 44 - Kale and chard.

120

Image 45 - My living border and hellstrip of marigolds, nasturtiums, and sweet potato vines.

121

GROWING FROM SEEDS

Seed packets are so cheap. Should you start your garden with those to save some money? What about those healthy-looking young plants in the garden section at the big box store? Is that a better option? Is that considered cheating?

No. It isn't *cheating*. Neither method is right, or wrong. How to start your garden is purely a matter of personal preference. Some people seem to feel like they can't consider themselves a real gardener if they buy young plants from the store. It is as if, in their minds, they are only a "real gardener" if they grow everything from seed. I personally think that has nothing to do with whether you are a real gardener or not. Anyone can buy plants from the store or chaos garden with scattered seeds. Heck, many times fruits and veggies will sprout at random from your compost without you doing a thing to it.

What makes you a gardener, at least in my eyes, is if you can keep them alive and get them to bear fruit when the time comes. However, even the most experienced gardeners will flub up with a crop or go out of town and forget to water. Or maybe hungry tomato hornworms decide to level your Roma tomatoes overnight, or a family of gophers decides to devastate your melon patch with a premature Thanksgiving feast. Or perhaps a random hurricane or storm snaps your stems, drowns your roots for days, or makes the conditions of your yard perfect for blight.

Unfortunately, when you garden, you have to prepare yourself for some things to die, whether it is because of your direct actions or things completely out of your control. The upside is, no matter how bad it gets, next year provides a clean slate where you can start anew, having learned from the mistakes of the year before.

So, back to how to start your garden! Some people are determined to grow everything by seed and then get really frustrated with themselves when some don't sprout or when they dampen off and wither away a few days after they spring up. This happens. Sometimes it is because you planted at the wrong depth. Sometimes it is because the temperature or timing wasn't right. Heck, sometimes it is even because a squirrel came after you planted your corn and picked your seeds out of the dirt and ate them before they even had a chance to emerge from their hard kernel shell. Despite the way my garden looks at the end of the summer and how much food I am giving away, I can promise you that all of the above have happened to me more times than I can count.

The garden's beginning is the part of your journey where you will need to grant yourself the most grace. If you are just starting out and you want to start with those pre-grown plants from your hardware store, do it! When you are first starting out, go with whatever method feels right for you.

Some people just want to master keeping things *alive* before adding on the challenge of growing things from scratch. And even though it doesn't feel like it once you walk away from the register, you're still able to make a food return on that initial investment beyond what you could buy in the produce department.

I don't judge anyone who buys starter plants from the store. Heck, I've had to do it myself with some plants that prove especially difficult to grow from seeds. For the life of me, I can't seem to grow cauliflower from a seed. Sometimes, with a small garden, it even makes more financial sense to do so.

Even if a pack of ten strawberry bare roots is $9, you might find a single pre-grown strawberry plant at the store for $5 that is almost guaranteed to fruit within a week or two, instead of a few weeks like any of the bare roots. The beauty of this is that after the strawberry plant is done producing fruit, it will spend the rest of the season creating runners. Along each runner, new leaves will form and, shortly after that, new roots, too. They reproduce and come back every season, so in year two, not only will you not have to buy another strawberry plant, but you might have six in total instead of one. And then twenty-four, the third year, depending on the fertility of your soil and the amount of compost they're getting. So by the second year alone, you are actually saving money by having bought a live plant, too.

Then again, if you planted your strawberries from bare roots, by year two, you might also have fifty strawberry plants if you get the runners to all root.

So which method is right for you? You decide!

GERMINATING SEEDLINGS INDOORS

If you have the room and the desire to start some of your garden plants and pollinator flowers inside from seeds, you can really get a massive jump-start on a shorter grow season in the north or prolong a longer one in the south.

You only need a few things to get started:

Seeds - This is the fun part. Make a list of your favorite foods or simply browse the winter and spring selection of seeds at your local garden center. Select a few seed packs that really appeal to you. I personally find no difference in the brands in my own experience. In fact, most of my yearly marigolds all came from one single $0.50 pack of seeds from the dollar store. For some fun and funky varieties that you can't find in the store, I recommend eBay and Gurneys.com

Image 46 - Seed packets from the garden center.

125

Labels - These become optional after a few years because you'll start to be able to recognize everything by the seedlings or leaves. They are so much fun to watch grow that it is hard not to notice all of their intricacies. However, when you are getting started, these can be a lifesaver. Not only will you remember what you planted where when you transplant your seedlings, you can ensure that you don't pair two plants that will clash (see the Companion Planting chapter coming up for a list of these), and also save yourself from plucking something you worked hard to grow because you forgot and thought it was a weed.

Image 47 - Plastic labels that can be re-used for many years. Simply mark them with a permanent marker and press the pointed end into the dirt.

A seed cell growing tray (or germination flat) - These are great and relatively cheap. They are readily available at pretty much any major store's garden department in the spring. These are typically plastic (so that they can be re-used over multiple seasons) with many cells, or they can be ones made of biodegradable materials with fewer cells that need to be re-purchased every year. No matter what you choose, I recommend having one with a transparent plastic lid to lock in the humidity to cut down on watering

126

while still letting light through. Seedlings seem to sprout best in consistently humid environments. Having this type of lid keeps the water and condensation in. Beneath your tray, I recommend having a heating mat to get seedlings to the right temperature to germinate indoors, and having this type of plastic lid keeps these tender babies from drying out and dying right off the bat.

Image 48 - Three plastic seedling trays in my south-facing kitchen window. The one on the far left has its lid on.

A seed mat - This is essentially a heating pad. Simply place it under your seedling tray and allow it to heat up the drained water and roots. Most of these are preset to certain temperatures that are ideal for the widest array of seed germinating. These are safe and effective, and I highly recommend them. For years, I was trying to get by without one of these by growing in windows that get lots of sun or setting them on a tepid radiator. As someone who has done a lot of experimenting with plants over the years, I will say that once I finally broke down and spent the $14 on one of these, I decided I was never going back to the old methods. I had a *much* higher rate of germination when using one. Almost all of my seedlings sprout every

127

time I use one of these seed mats. It's incredible.

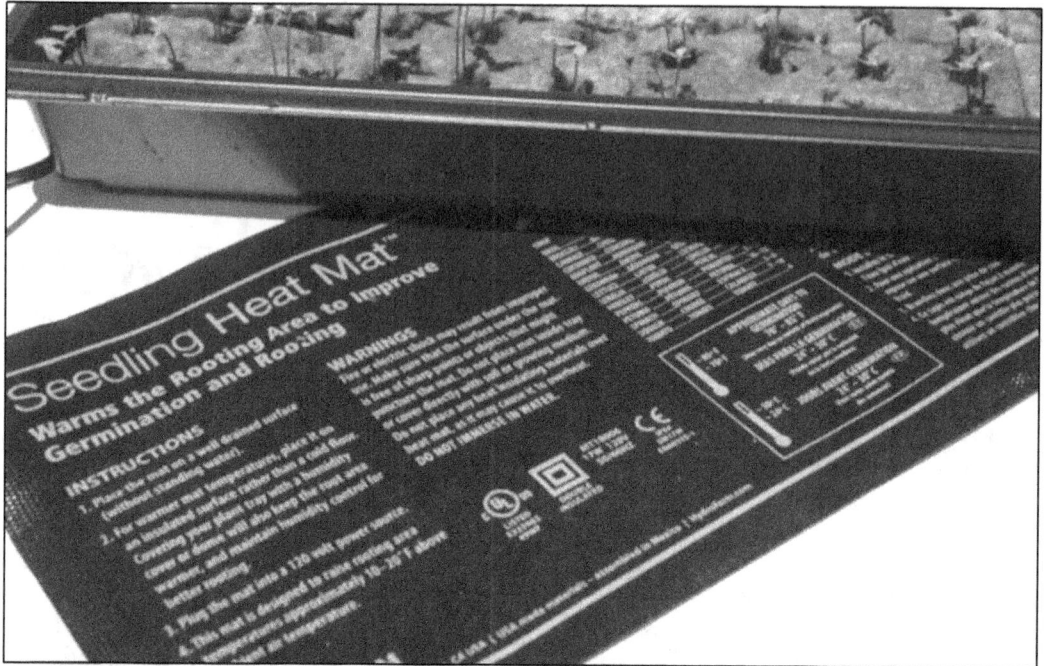

Image 49 - A seedling heat mat.

Loose dirt or a bag of seedling starter mix - You will need dirt to fill the cells of your germination flats. I recommend a loose blend containing either Perlite, peat moss, or both. You can mix your own at home with dirt from your yard mixed with those amendments. However, I advise you, especially when starting out, to spend the extra $5 to just buy a bag of seed-starter dirt at the store. They carry it in the same aisle you buy your seed flats/cells.

I recommend it because most commercially bagged seed starter mixes have been baked at high temperatures, killing off a lot of the soil-borne diseases and fungi that can lead to damping off, which is what kills a lot of sprouted seedlings. Often, the bagged seed starter soil you buy at the store already contains Perlite and peat moss, too, so you're giving your plant the best shot at survival and healthy growth right off the bat.

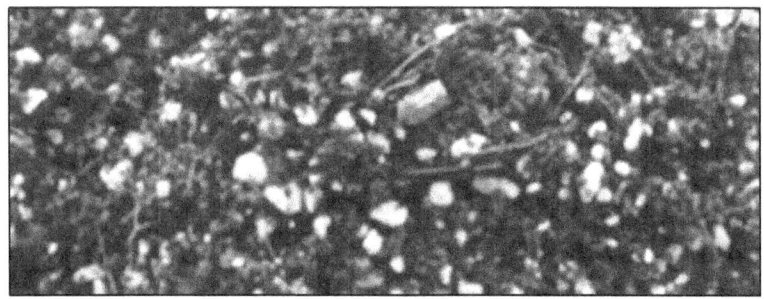
Image 50 - A good seedling starter soil is loose and soft.

LED lights or a sunny spot for placement - This is crucial. Plants need light. They use photosynthesis, which is where light energy converts water and carbon dioxide into the sugars and fuel needed for healthy growth. Chlorophyll (the leaves' green pigment) turns sunlight into a form of energy for the plant.

Plants can't create their own food or thrive without light. So to get those little babies off to a strong start, you either need to place them in a sunny spot (like a south-facing window) or get them some supplemental light, like an LED grow light.

These low-cost, high-tech LED lights now come in a variety of shapes and sizes, and, unlike the old days, they draw very little power. They can clip onto pots or shelves and bend in various ways. They even have multiple modes and timers on them so that you can plug them in, adjust your settings, and allow the light to do the rest until they're ready to go outside in a few weeks!

Alternatively, there are full-spectrum LED bulbs now that have options like full-spectrum white, red, and blue modes. For seedlings, I recommend either full-spectrum white or blue, as red is more for encouraging established plants to fruit. Blue and full-spectrum white (since white has the blue spectrum *in* it) both encourage foliage.

Image 51 - one of my full-spectrum white bendable LED grow lights clipped to my lemon tree's pot, hovering over two avocado trees and a hibiscus in the corner of a room.

Patience, experimentation, and forgiveness - This may sound like something silly or abstract that a life coach might say, but I really mean it.

➤ Patience is key to learning how to grow bigger and better hauls each year. By not giving up, you ensure that eventually you will probably get it right. Gardening should be a relaxing venture, not a deeply frustrating one, but mistakes will be made, and every year, it will get a little easier as you gain knowledge and experience. Don't give up if you feel like you're not good at it immediately. Almost none of us gardeners ever are, I assure you.

➤ Experimentation keeps things fresh and fun. Growing new strains and varieties can help you find plants you really love to watch grow. This is exactly how I developed my obsession with lemon cucumbers and purple kale, two absolute staples of my garden from now on. I found them while browsing some eBay seller stores and figured I'd experiment a little and try them out. They are now some of my favorite things to watch grow, and they're delicious!

➢ Forgiveness is crucial because you *will* make mistakes. I know I've said it before, but it is worth repeating. Being able to forgive yourself when you weed-whack the vine of your gorgeous winter squash or when your beefsteak tomatoes start to split will help keep you from quitting the entire venture before you really blossom as a well-rounded gardener. If you remember, in the intro, I told my story about the Black Walnut tree in my yard. If I hadn't forgiven myself for killing so many plants or had the patience to keep trying after those first two terrible seasons, I never would have learned that I was not actually the problem at all and that my *location* and *garden type* were the real things thwarting my success. So give yourself some grace when starting your gardening journey. I promise, if you stick with it, there's a garden veggie master within you.

Once you have acquired your supplies, fill your cells with moist dirt and plant the seeds according to the instructions on the packet or website. A general rule of thumb with planting seeds is to plant a seed twice as deep as it is wide. So if it is only a millimeter wide, plant it roughly two millimeters beneath the soil's surface. Also, there are some seeds you won't even need to sow at all. You can simply sprinkle them on the dirt in your grow cell. This is usually how I grow my amaranth, coleus, and poppies.

Do not press the dirt down on top of your seeds. Too much compaction can mean your seed sprouts, but it isn't strong enough to break through. It eventually dies there. Likewise, even if the plant makes it out of the top of the soil, if the dirt is too compacted, it will stunt the root growth and make for a stunted plant. Keeping the soil a little loose ensures the seedling can break through to that much-needed sun and the roots can seek their way down to the water well at the bottom to drink.

Make sure you label everything with either plastic labels or a key diagram as you go, so you know what you've planted.

Plant two to three seeds in each seed cell to up your chances of successful germination. If they all grow, simply pinch off the head of the one(s) that don't look the strongest or healthiest and keep the cream of the crop.

Set them on your grow mat, give them a little water and light, put the lid on, and watch them grow! I start most of my garden veggie seeds in mid-late February so that they can grow about eight weeks before I put them in the ground. You can always grow them earlier or later, depending on your needs and growing zone. You also don't have to grow them for eight weeks like I do, either.

After your seedlings have grown some nice roots and your last frost has passed, it is time to harden off your tender seedlings for a few weeks before putting them in your in-ground garden, raised bed, hanging planters, or containers.

Note: Waiting until after the last frost to transplant your seedlings is absolutely crucial. *Always wait if you're unsure.* I cannot even begin to tell you how many times I've gotten overzealous and transplanted gorgeous, healthy seedlings into my in-ground garden in the spring and watched two inches of early May snow pile on them a week or two later and kill them all in a single morning. It is devastating, though not irreparable. You can always direct-sow new, fresh seeds if this happens. It just means that your plants will be a few weeks behind by the time summer kicks in, or they'll have a few weeks less of that glorious harvest time at the end of the season. It's a bummer, but it's also not the end of your grow season if it happens.

Hardening off your seedlings just means that you are acclimating your plants to outdoor conditions over the course of a week and a half or so. During this time, you can take your seedling flats and place them outside for a few hours, preferably starting in a spot out of the wind and in indirect light (not straight in the blazing sun or they'll burn, just like a child). Then, each day, they can gradually spend more time outside and in spots that have gradually more wind and sun.

If you are a patio gardener or someone who only has a tiny swath of available yard space, this may not apply to you. You may just have to put them right atop your garden and increase the amount of time they are there each day.

I personally sometimes skip the hardening off step altogether because now I know some plants do just fine without it, but I always recommend doing it if you're a newer gardener, because you *will* have the best chance for success by hardening off your seedlings.

When transplanting your seedlings, I always recommend doing so on the early morning of a cloudy day or just before sunset if there are no cloudy days in your forecast. This gives the plant a chance to start rooting itself and acclimating without experiencing the added trauma of being in the brutal full sun on its first day in the ground. Plants typically bounce back and start looking great again much faster if you do this.

DIRECT SOW

Direct sowing is when you plant your seeds directly in your garden bed, hanging planters, or containers after the last frost. You can direct-sow most things. The downside is that, if you have a shorter growing season, like I do, you cut down on several extra weeks of harvesting at the end of your season because your plants are weeks behind those that were started inside.

That said, many plants *should* be planted using the direct sow method because they either do not transplant well or do poorly in grow flats.

Here are some crops that, in my experience, do much better when sown directly in your garden or raised bed after your season's last frost:

- Corn
- Artichokes
- Snap peas
- Beets
- Potatoes
- Sweet potatoes
- Green beans
- Asparagus
- Borage
- Chard
- Poppies
- Kohlrabi
- Onions
- Marshmallow
- Morning glory

- Coneflower (echinacea)
- Sunflowers
- Carrots
- Catnip
- Spinach
- Lettuce
- Mint
- Zinnias
- Marigolds
- Nasturtiums
- Strawberries
- Sweet Peas

Conversely, here is a list of plants I recommend starting indoors:

- Dill
- Borage
- Nosturtium
- Poblanos
- Jalepenos
- Habaneros
- Banana peppers
- Bell peppers
- Scorpion peppers
- Chili peppers
- Tomatoes
- Cherry tomatoes
- Cucumbers

- Honeydew
- Watermelon
- Squash
- Eggplant
- Broccoli
- Cauliflower
- Basil

MINT & OTHER INSIDIOUS INVASIVES

It is worth mentioning that some plants, when sown or transplanted directly into the ground or a raised bed, can spread aggressively and should, in my opinion, only be potted or planted in hellstrips where concrete or other materials will limit their spread. Mint, for example, is one of them.

Now, you might be thinking, I love mint! I want to have a lot of it. Why shouldn't I put some in my garden?

Some of these (often beautiful) plants, like spearmint, wisteria, or lemon balm, spread through underground runners instead of seeds. This makes them almost impossible to eradicate once their veiny rhizomes have spread through your garden. You might still be thinking, *Okay, what's the big deal? An effortless plant that I love in abundance sounds like a good thing.*

Well, these perennial or aggressively re-seeding plants, once they spread, will out-compete all of your other fruits and veggies for precious nutrients and water. Their underground root system can even choke out your annuals beneath the soil by leaving very little room for your other plants to spread their roots. When this happens, they will suffer severely or even die.

These runners don't stay contained where you think they will. They will spread through cracks and under fences, and soon your neighbors will be plagued by your choice to plant such an aggressive species anywhere in their vicinity.

In a few years, it will be *their* neighbor's problem, too.

The solution to this is simple: you can still have these plants and enjoy their benefits and beauty by keeping them in containers. This way their root system stays... You guessed it... *contained,* ensuring that it won't spread to everything around it.

Image 52 - A side view of mint's vein-like rhizomes.

You might hear the word invasive thrown around, too. Sometimes this term gets misused, as invasives often vary by region. These are simply non-native plants that reproduce rapidly, outperforming native plants and causing harm to the environment, human health, animal habitats, or the economy.

Weather and location matter, too. A plant that is highly invasive to me here in Connecticut might not be invasive at all to someone in California,

Canada, or Northern Ireland. A Google search will usually reveal a variety of resources about what plants are invasive in your area.

Japanese honeysuckle is a great example of this. It is a native plant in Japan (as the name implies) that is very invasive in North America and should not be planted in the ground here, despite being sold in a wide variety of online and big box stores.

Native species provide great habitats and food sources for native birds, bugs, and a slew of other animals. These are also naturally pest and disease resistant and they typically don't require nearly as much water as non-natives. They can enrich your area in a myriad of ways beyond just beauty.

Here is a list of some North American natives:

- Southern magnolia tree
- Douglas fir
- Sugar maple
- Bald cypress
- Agave
- Common milkweed
- Coral honeysuckle
- Wild blue phlox
- Asters
- Wild bergamot
- Black-eyed Susans
- Purple coneflower
- Indiangrass
- Big bluestem
- Highbush blueberry

- Corn
- Squash
- Raspberries

Here are a list of some fast-spreading plants to avoid planting outside of a container in North America:

- Spearmint
- Rhubarb
- Catnip / Cat mint
- Bee balm
- Chocolate mint
- Wisteria
- Bamboo
- Lemon balm
- Lamb's ear
- Virginia creeper
- Yellow iris
- Hemlock
- Russian olive
- English ivy
- Garlic mustard
- Japanese honeysuckle
- Cape honeysuckle
- Autumn olive
- Purple loosestrife
- Lantana
- Burning bush
- Japanese barberry

Have you ever heard the comedic expression, "Kill it with fire?" Well, some of the above plants, you can't even really do *that* with. Once established, no amount of weeding in the world might ever *fully* rid you of things in your ground, especially bamboo, kudzu, wisteria, tree of heaven, and mint. The best course of action is to learn to identify them, keep them regularly shorn to the ground, and pull up as much of the root system as you have the energy for.

Image 53 - Left, kudzu. Right, black swallow wort.

In my neck of the woods, there are poisonous nightshades that spread like wildfire and a plant called black swallow wort that is the bane of my existence (the swallow wort, *especially*, because it poisons monarchs and spreads through mint-like rhizomes). Anything shy of ripping up the soil several feet down and plucking out every last vein-like runner means I'm stuck with this stuff. I just have to keep them pulled as they emerge so they don't poison the butterflies and spread via their prolific seed pods. Invasives happen, but we don't need to exacerbate them either by planting more.

It is all too tempting to plant things like mint in the ground, but once they're there, you've essentially signed a long-term contract.

If you're still dead-set on planting mint or wisteria in your in-ground garden, I encourage you to envision yourself in ten years, digging it out of the ground with sweat pouring down your face, cursing at your ten years younger self. If that visual doesn't deter you, then by all means, plant it.

Image 54 - Top: Tree of heaven. Bottom Left: A yard destroyed by bamboo. Bottom Right: English ivy and bamboo invading woodlands. Photos courtesy of the North Carolina Native Wildflower Society.

COMPANION PLANTING

Here's the fun part. Now that you have decided to make (or improve upon) this year's garden, it is time for my secret weapon. These two words will change your gardening journey in a positive way forever. What are they? As you could surmise from the heading of this chapter:

Companion planting.

These two words forever changed the way I garden, and when *anyone* asks, "What's your secret?" or "How'd you get such a green thumb?" Those two words are my answer, usually followed by about thirty minutes of me nerding out and telling the person so many little tips and facts about it that their eyes glaze over.

What is companion planting? Put simply, it is the pairing of two or more beneficial plants in your garden. The benefits may be anything from providing structure or shade, to even something you can't see with your eyes, like one plant affixing nitrogen to your soil that the other plants around it can use for fuel.

You read that right. In some cases, one plant can literally feed another while both are growing you some delicious vegetables to harvest later.

I discovered this almost accidentally about four years after I started gardening. Every year prior, I had planted my garlic in one measured sixteen-inch square. Next to it, I planted the same size swath of yellow onions. Just past that, my carrots. Every year, my carrots grew alright, but the onions and garlic stayed bare patches of dirt. Year after year, it continued to baffle me. It wasn't until I started searching the internet, on a quest for answers, that I found out that garlic doesn't grow well alongside other alliums (like onions).

Aha! Mystery solved.

After a moment of smug satisfaction for having gotten to the bottom of the issue, I was posed with a new burning question:

Okay, if I can't plant onions and garlic together, what *can* I plant them alongside?

Almost immediately, I ended up on a website about companion planting. The concept was so simple, I honestly felt pretty dumb for not realizing it was even a "thing" sooner.

There is a whole list of upsides to companion planting, but one of the greatest perks of them all is the amount of *space* it saves when you companion plant. In some cases, you can plant things right next to each other, and instead of causing overcrowding, like you'd think, they prevent pests or diseases.

Here's a great example of a symbiotic companion planting relationship:

Tomatoes, basil, and marigolds. A magical trio.

My tomatoes did poorly the first few years I planted them. Then, when I discovered companion planting, I was doubling and tripling my tomato harvests. Why? Because I was planting all three of those crops all in the same exact space that I'd normally plant *just* my tomatoes.

So... why did that *work*? Because basil and marigolds repel a whole slew of pesky insects like aphids and tomato hornworms (all while attracting pollinators like bees in droves). Basil also simultaneously improves the tomatoes' flavor. To top it all off, the marigolds and basil leaf out over the dirt near the base of the tomato vine, and in heavy rains, they often prevent mud from splashing up onto the leaves of the tomato plant, thus cutting back on a variety of soil-borne disease pathogens. Then, when it is time to harvest, I have so much basil that I can make pesto or one killer tomato sauce infused with basil!

Here's another magical trio for you:

It is called the Three Sisters. These consist of green beans, squash, and corn. The corn provides a natural trellis for the climbing beans. The beans fix nitrogen into the soil (instead of taking it *from* the soil like most plants as a source of fuel), which fuels the squash and corn, both heavy feeders. The squash spreads along the ground and shades the roots of the corn, keeps moisture from evaporating too quickly, and deters weeds for all three crops.

That's pretty incredible if you ask me.

Most plants have symbiotic companions, and many have what I like to pretend are enemies. Your garden is Manhattan's Upper West Side, and you've got your Jets and your Sharks in an ongoing, active turf war in your garden like West Side Story. So pair your Jets and your Sharks in their respective groupings, Officer Krupke. That way, no one has to end up in a rumble with fatalities. *Comprende*?

So what plants go well together and what causes the aforementioned *rumble*?

I'm so glad you asked! I have spent a few years compiling some information about some common crops that will help you. Not everything will be laid out for you here (if so, this book would be 900 pages long), but the best news is, any plant you want to know the companions and enemies of is only a twenty-second internet search away.

Still, I feel the need to save you a whole heap of time by sharing with you my notes about some of the most common crops you will find in the average home garden...

PLANTING PALS (& ENEMIES)

CUCUMBERS

These pair well with garlic, tomatoes, okra, marigolds, basil, artichokes, corn, borage, and nasturtiums.

Avoid planting them with potatoes, melons, squash, sage, and rosemary.

PEAS

Peas don't like a lot of water. They pair well with tomatoes, okra, artichokes, lettuce, and peppers.

Avoid planting them with alliums like onions, chives, garlic, and leeks.

CARROTS

Pairs well with beans, onions, asparagus, peppers, rosemary, sage, chives, amaranth, lettuce, and tomatoes.

Avoid planting them with dill, fennel, parsley, potatoes, celery, & parsnips.

POTATOES

Potatoes pair well with cabbage, corn, beans, horseradish, asparagus, lettuce, spinach, basil, parsley, petunias, alyssum, nasturtiums, and marigolds.

Avoid planting them with tomatoes, peppers, eggplant, carrots, beets, turnips, cucumbers, raspberries, squash, or pumpkins.

BEANS

They pair well with carrots, corn, lettuce, tomatoes, potatoes, and squash.

Avoid planting them with onions, garlic, leeks, fennel, sunflowers, peppers, chives, cucumbers, or beets.

ONIONS AND CHIVES

They like carrots, strawberries, tomatoes, broccoli, cabbage, peppers, lettuce, dill, celery, chard, and marigolds.

Avoid planting them with beans, peas, sage, asparagus, leeks, shallots, or garlic.

BEETS

They pair well with cabbage, lettuce, and garlic.

Avoid planting them with pole beans, spinach, chard, corn, potatoes, tomatoes, and eggplants.

TOMATOES

They love basil, nasturtiums, marigolds, asparagus, onions, chives, carrots, and peas.

Avoid planting them with broccoli, cabbage, cauliflower, potatoes, peppers, eggplant, and corn.

CABBAGE

They pair well with potatoes, onions, chives, and artichokes.

Avoid planting them with radishes, tomatoes, broccoli, cauliflower, kale, strawberries, pole beans, and dill.

RADISHES

They pair well with artichokes, peas, lettuce, carrots, onions, cucumbers, and beans.

Avoid planting them with cabbage, broccoli, cauliflower, turnips, hyssop, and potatoes.

BASIL

They pair well with tomatoes, peppers, asparagus, marigolds, broccoli, tomatoes, potatoes, and okra.

Avoid planting them with fennel, sage, cucumbers, thyme, and rosemary.

MARIGOLDS

They pair well with tomatoes, peppers, basil, cucumbers, asparagus, onions, potatoes, and lettuce.

They repel tomato hornworms, nematodes, and deer.

Avoid planting them with beans, other legumes, broccoli, and cabbage. They can also be planted as a fantastic cover crop for potatoes to repel nematodes, but shouldn't be planted simultaneously with them.

CUCUMBERS

They pair well with broccoli, okra, peppers, beans, peas, corn, radishes, dill, and marigolds.

Avoid planting them with watermelons, cantaloupe, honeydew, pumpkins, banana melons, potatoes, sage, and mint.

PEPPERS

They pair well with onions, chives, basil, oregano, thyme, nasturtiums, borage, parsley, rosemary, dill, carrots, cucumbers, squash, spinach, lettuce, chard, corn (if the corn acts as a wind and sun barrier), peas, and marigolds.

Avoid planting them with cabbage, kale, cauliflower, beans, peas, eggplants, tomatoes, broccoli, potatoes, and fennel.

LETTUCE AND SPINACH

They pair well with tomatoes, potatoes, asparagus, okra, onions, chives, carrots, cilantro, dill, peppers, mint (if planted in containers), garlic, beans, beets, broccoli, corn, peas, radishes, and marigolds.

Avoid planting them with broccoli, kale, cauliflower, fennel, celery, and parsley.

STRAWBERRIES

They pair well with asparagus, onions, chives, borage, thyme, sage, spinach, lettuce, marigolds, alyssum, beans, and peas.

Avoid planting them with broccoli, tomatoes, potatoes, peppers, eggplants, cucumbers, watermelons, pumpkins, honeydew, and cantaloupe.

ASPARAGUS

They pair well with strawberries, tomatoes, lettuce, carrots, basil, spinach, parsley, dill, potatoes, marigolds, nasturtiums, cilantro, and rhubarb.

Avoid planting them with onions, chives, garlic, leeks, and shallots.

SUNFLOWERS

They pair well with dill, corn, squash, onions, zinnias, marigolds, and cucumbers.

Avoid planting them with potatoes, rhubarb, beans, or within three feet of tomatoes.

DILL

They pair well with onions, chard, peppers, corn, cucumbers, lettuce, asparagus, cabbage, broccoli, sunflowers, and cauliflower.

Avoid planting them with tomatoes, potatoes, carrots, eggplant, celery, or lavender.

EGGPLANT

They pair well with peppers, oregano, marigolds, nasturtiums, spinach, borage, beans, mint (in containers), thyme, and basil.

Avoid planting them with cucumbers, corn, pumpkins, melons, potatoes, peppers, fennel, geraniums, broccoli, and tomatoes.

GARLIC

They pair well with tomatoes, cabbage, kale, broccoli, carrots, potatoes, spinach, lettuce, peppers, strawberries, chamomile, and nasturtiums.
Avoid planting them with onions, chives, leeks, shallots, beans, peas, sage,

asparagus, and parsley.

SWISS CHARD

They pair well with broccoli, onions, chives, beans, peas, cabbage, kale, cauliflower, garlic, radishes, cilantro, and marigolds.

Avoid planting them with potatoes, corn, beets, cucumbers, spinach, watermelons, honeydew, cantaloupe, and banana melons.

WATERMELON, HONEYDEW, AND CANTALOUPE

They pair well with okra, marigolds, basil, mint (in a container), garlic, onions, beans, and peas. Melons are heavy feeders with long vines. Avoid overcrowding these.

Avoid planting them with cucumbers, pumpkins, squash, potatoes, tomatoes, sunflowers, and aster.

KALE

They pair well with dill, rosemary, sage, beans, peas, marigolds, radishes, basil, chives, mint (in containers), thyme, and onions.

Avoid planting them with broccoli, cauliflower, cabbage, tomatoes, strawberries, cucumbers, Brussels sprouts, radishes, and corn.

OKRA

They pair well with cucumbers (as long as they aren't too close), watermelons, cantaloupe, sunflowers, radishes, basil, cosmos, zinnias, calendula, lettuce, and peas.
Avoid planting them with squash, fennel, tomatoes, eggplant, cauliflower,

151

broccoli, and sweet potatoes.

CORN

They pair well with squash, beans, peas, potatoes, lettuce, sprawling melons, marigolds, and peppers (if the corn is acting as a wind and sun barrier).

Avoid planting them with tomatoes, eggplant, broccoli, cabbage, fennel, cauliflower, kale, kohlrabi, and Brussels sprouts.

SQUASH/ZUCCHINI

They pair well with beans, sunflowers, corn, peas, mint or catnip (in containers), tomatoes, nasturtiums (they prevent vine borers), peppermint, dill, basil, peppers, oregano, lemon balm, and borage.

Avoid planting them with okra, melons, cucumbers, pumpkins, potatoes, fennel, cabbage, and beets.

BROCCOLI AND CAULIFLOWER

Spread these out. If grown too close together, these brassicas will compete for nutrients. They pair well with chard, lettuce, spinach, dill, rosemary, basil, mint, nasturtiums, marigolds, sage, and cucumbers.

Avoid planting them with tomatoes, eggplant, peppers, beans, squash, strawberries, kale, and corn.

ARTICHOKES

They pair well with cabbage, cucumbers, radishes, peas, asparagus, marigolds, thyme, broccoli, and sunflowers.
Avoid planting them with fennel and potatoes.

GRAPES AND MUSCADINES

Avoid highly fertile soil. This will make grapes grow too fast and focus on foliage growth over fruiting. They pair well with peas, basil, oregano, rosemary, chives, beans, and clover.

Avoid planting them with anything that shades the ground where their roots are. They like warm roots. Avoid neighbors like cabbage, fennel, and radishes.

Image 55 - Green grapes on a rabbit wire trellis

153

SUCCESSION PLANTING

Succession planting is a fantastic technique used to maximize your grow season to the fullest by staggering the planting and harvesting of several sets of crops in intervals within the same season in a continuous way instead of getting a massive harvest all at once. You can alternate these crops or plant the same plants all over again.

Why would anyone do this, you might ask? It sounds like twice the work.

Technically, you're right. However, there are many instances in which one might do this. It is a great technique that can make the most out of a tiny garden space or time your harvests so that your food is always fresh.

Here's an example. I love salads (obviously!) In the regular method of planting, I head out to my garden in the spring, transplant all fifty of my butter-crunch lettuce seedlings, and call it a day. In a few weeks, I am in salad overload, scrambling to make salad after salad with all of the mature lettuce before it all bolts and starts to taste awfully bitter. Then, the plants bolt, and I say, "That was fun. I guess that's all the homegrown salads I get until next spring."

But there's still so much of the grow season left. If I stagger my planting and plant some of my seeds now, in a few weeks, I can have some fresh salads, yank those bolted crops, put them in the compost, and then plant some more seeds the same day. In a few weeks, I have a whole new, fresh batch of salads. This can go on several times in a single grow season.

In my grow zone, there are charts that recommend what to start indoors, what to plant each month, what to replant in other months, and when to harvest each.

If you'd like to try succession planting, simply find a reputable planting schedule for your grow zone and try it out! I learned to grow so much more by taking advantage of these successive planting opportunities.

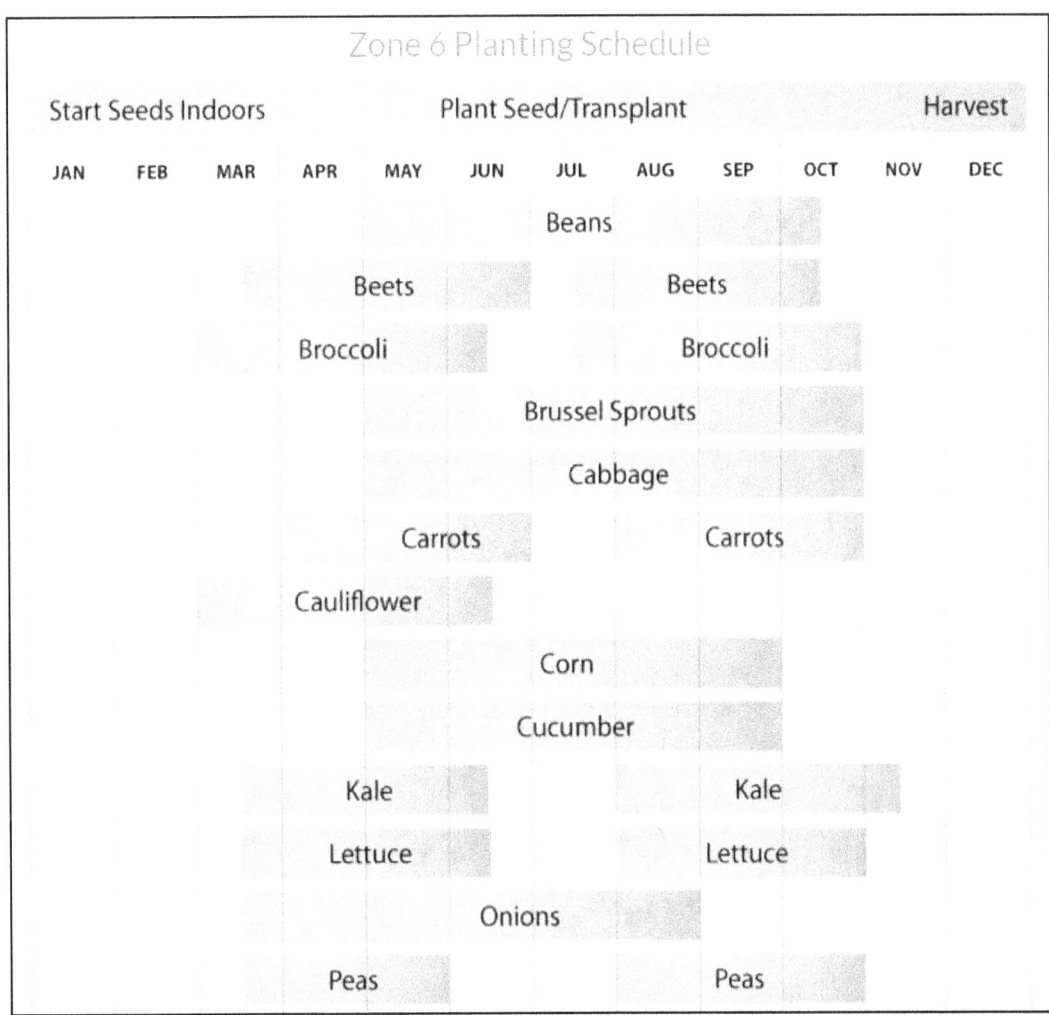

Image 56 - A succession planting guide for my grow zone. Photo courtesy of Urban Farmer. www.ufseeds.com

THE FLOOD: WHAT PLANTS WILL THRIVE IF YOU'RE AN OVER-WATERER

I am guilty of over-watering frequently. I just get so excited and overzealous, I can't help myself. Plus, it gives me an excuse to inspect, enjoy, and keep an eye on all my plant babies.

If you are like me, you might benefit from knowing what plants will thrive if you like to water a lot, like I do.

Please note that, while these plants are thirsty, it is important to water them correctly, which is at the base of the plant, not the leaves. Leaving the leaves of any plant wet for a long time invites a variety of molds, mildews, and other fungal infections, which can kill your plants. So even though your pumpkins adore a big, long drink of H2O, if you give their leaves mildew by hosing the whole plant down, you may watch it die a slow death.

Water at the base, where the plant emerges from the dirt, and keep those leaves as dry as possible. This is less true for strawberries, though, as their leaves are smaller and dry more quickly.

And remember, if you start seeing the sudden appearance of mushrooms in your garden bed, you're probably overdoing it with the watering.

Here is a list of thirsty plants that, in my experience, will gladly accept all that extra attention with the hose:

- Strawberries
- Pumpkins

- Raspberries
- Blackberries
- Cucumbers
- Watermelons
- Banana melons
- Corn
- Blueberries
- Zucchini
- Butternut squash
- Eggplant
- Honeydew
- Cantaloupe
- Crookneck squash

Image 57 - (Top) A week's worth of mid-summer zucchinis. (Bottom) Two varieties of eggplants I grow in my garden.

157

Image 58 - (Top) A bowl of fresh strawberries. (Bottom) One of my first bush baby watermelons. They're compact, delicious, and great for a small garden.

Image 59 - (Top) A tub full of blackberries straight off the vine. (Bottom) Highbush blueberries ripening on the plant.

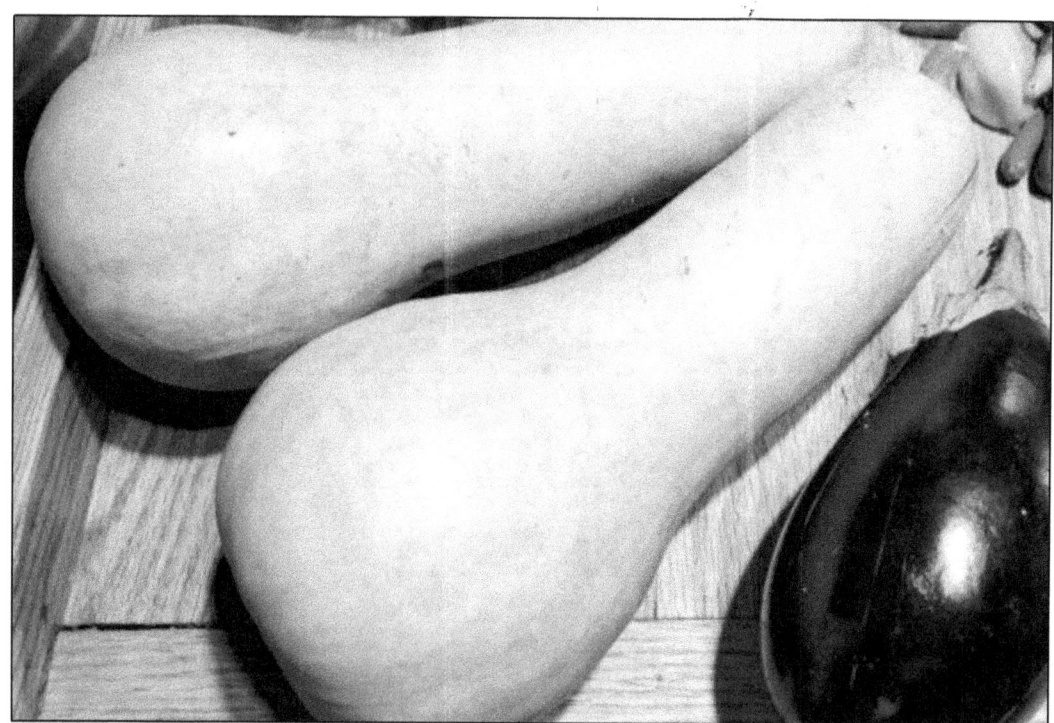

Image 60 - (Top) One of my garden favorites, lemon cucumbers. (Bottom) Butternut Squash and an eggplant.

160

THE DROUGHT: WHAT PLANTS WILL THRIVE IF YOU'RE AN UNDER-WATERER

Okay, maybe you're like one of the billions of other people out there who don't like watering as much as I do. Maybe you don't have the time to stand out in the yard three or four times a week like I do. Don't worry, there are lots of crops that are ideal for your situation. By growing more of these, you will be able to reap large harvests with little effort.

Here are some plants that thrive with neglect, at least watering-wise:

- Tomatoes
- Grapes
- Herbs
- Olives
- Carrots
- Potatoes
- Sweet potatoes
- Peppers
- Figs
- Lavendar
- Agave
- Coneflower
- Sage
- Purslane
- Black-eyed Susans

Image 61 - A variety of tomatoes from my garden.

Image 62 - (Top) Jalapenos, habaneros, hot banana peppers, etc. (Bottom) a netted bag of potatoes harvested in the fall.

163

SHADE & COOL WEATHER PLANTS

- Kale
- Parsley
- Cilantro
- Lettuce
- Cabbage
- Winter squash
- Garlic
- Arugula
- Mustard greens
- Scallions
- Beets
- Turnips
- Spinach
- Carrots
- Chard
- Bok choy

164

FULL SUN PLANTS

Full sun is preferred by many fruits and vegetables that produce flowers, fruit, or roots. Here is a list of some garden sun-lovers:

- Tomatoes
- Peppers
- Cucumbers
- Zucchini
- Squash
- Melons
- Pumpkins
- Corn
- Broccoli
- Eggplant
- Okra
- Potatoes
- Strawberries
- Lemons
- Limes
- Oranges
- Pomegranates
- Figs
- Pineapples
- Green beans
- Pole beans
- Carrots
- Onions
- Tomatillos

Image 63 - (Top) Cherry and pear tomatoes. (Bottom) Four varieties of eggplant, banana peppers, broccoli, and three varieties of tomatoes.

POLLINATOR CROPS

By now, you know that I feel *very* strongly about planting natives and pollinator crops. It doesn't matter if you are the manliest of men or are a woman with the tiniest of garden spaces; I recommend adding pollinator crops and flowers to your garden.

This is one of the biggest weapons in your arsenal for improving your yields.

You could have ten sprawling acres of land to garden on, but if the pollinator insects aren't hanging around and rolling in all of that powdery goodness, you're going to have an underwhelming yield come harvest time.

By adding pollinator flowers to your in-ground or patio garden, raised beds, hanging planters, and/or containers, you are *vastly* improving the likelihood that even your small suburban garden will produce for you in a *big* way.

This is *precisely* why this book is called what it is.

Garden *size* doesn't matter as much as you think it might. It's what you *do with that space* that makes all the difference.

Pollinator flowers are crucial. They provide a food source for the insects that will make those flowering crops turn into fruiting ones. They are necessary for the reproduction of our fruits and veggies. About one-third of all the produce the world eats requires pollination and unless we want a future where we are all doing it by hand, standing outside in our pajamas with a paintbrush stroking the bristles across each plant like we're Edvard

Munch or something, then we need to feed and provide a hospitable environment for all those beautiful, beneficial, buzzing, flitting, crawling, fluttering creatures.

So what exactly *are* our pollinators?

Pollinators can be a lot of things. Some of the most common ones are:

- Bees
- Butterflies
- Ladybugs
- Hummingbirds
- Bats
- Moths
- Flies
- Beetles, etc.

Even some small *rodents* are pollinators, believe it or not.

Pollinator flowers and crops increase biodiversity and promote strong, healthy, stable ecosystems in nature by supporting the entire food web, from the animals to the plants themselves.

Flowers and crops that draw in pollinators:

- Blanket flower
- Milkweed
- Black-eyed Susans
- Coneflowers
- Sunflowers
- Marigolds
- Cornflower
- Candytuft
- Cleome
- Nasturtiums
- Verbena
- Calendula
- Bee Balm
- Goldenrod
- Cosmos
- Clover
- Yarrow
- Joe-Pye weed
- Blazing star
- Anise hyssop
- Dill
- Foxglove
- Smooth spiderwort
- Eastern columbine
- White beardtongue
- Spotted crane's bill

Image 64 - (Prior Page) A bee feasting on pollen. (Top) - A butterfly on Purple Coneflowers. (Bottom) My fence of Morning Glories.

169

- Turk's cap lily
- Violets
- New England aster
- Mountain mint
- Bird's foot violet
- Carolina rose
- Common witch hazel
- Parsley
- Blue vervain
- Virginia bluebells
- Blue false indigo
- Alyssum
- Pussywillow
- Black chokeberry
- Highbush blueberry
- Gay butterfly weed
- Cardinal flower
- Salvia
- Passionfruit flower
- Daffodils
- Dandelions
- Morning glory
- Zinnia
- Borage
- Golden groundsel
- Woodland phlox
- Petunia
- Brown-eyed Susans

Image 65 - (Top) Bee Balm. (Bottom) Foxglove.

170

- Wild bergamot
- Tulips
- Purple prairie clover
- *...and many more!*

Image 66 - Cosmos. A large, stately plant that blooms in the fall. Pollinators love them.

Image 67 - Purple passionfruit flower. While these are vigorous and aggressive growers and should be kept in containers, pollinators love them and they also produce edible passionfruits.

Image 68 - One of my tulips. I adore these early spring pollinator flowers. They're stunning and they let me know that spring has truly arrived.

WORTH THEIR WEIGHT IN GOLD: MARIGOLDS, BASIL, AND NASTURTIUMS

Personally, I can't get enough of three things in my garden: marigolds, basil, and nasturtiums.

Not only do they look stunning, but they are all-natural pest repellents or trap crops, which means they draw in bees, ladybugs, and butterflies while warding away aphids, tomato hornworms, and nematodes. They suppress weeds and improve soil health. Their seed pods are easy to harvest, so that you have free crops for the next year and, best of all, they can go in your compost bin at the end of your grow season to become food for next year!

Not to mention, I think basil is delicious, and fresh basil can be outrageous in the store. Between those and my cherry tomatoes, I have summers full of *caprese* for next to nothing. In today's day and age, with the prices at the supermarket, it honestly makes me feel like royalty.

Might as well be beluga caviar.

Nasturtiums are what is called a trap crop. I mentioned this earlier in the section on borders, but it is a simple concept. These are attractive plants (in this case, one with a myriad of benefits) that draw the attention of bugs and critters away from your cash crops (aka the things you're growing for yourself).

Nasturtiums are really stunning. They even come with variegated leaves sometimes. They have gorgeous flowers on the warm side of the spectrum (red, orange, and yellow) and brilliant sand-dollar-esque leaves. Personally, I think they smell lovely. They have a very distinct, almost peppery-floral scent. Their flowers really pop among all the green in the garden, and I get questions and compliments on them all the time.

They vine out across the ground, mound like a fountain, and can even climb fences (though not with the stranglehold of something like a morning glory, cold-hardy kiwi, or grapevine). The bees, ladybugs, and lacewings love them, and they repel those pesky aphids and white flies. They, too, suppress weeds and break up compacted soil where they are planted.

They're also extremely easy to start from seed and keep alive. For me, they are almost an effortless crop despite all of its many functions.

Another added benefit of nasturtiums is that they're edible! The leaf, the flower... all of it. They taste peppery and flavorful, even a little spicy. They are rich in vitamin C. They also contain vitamins A and K, and the flowers themselves have B vitamins and beta carotene, too. They go great in salads or sauteed with other veggies. I put them in my stir fry.

Basil, along with being delicious, repels a boatload of damaging insects. But when they go to seed, the bees and butterflies have a field day! In addition to being an edible, chemical-free pest deterrent, they can also enhance the flavor of the things they are planted next to, like tomatoes.

There are multiple varieties available, many of which taste very different. I personally grow four separate types in my garden since they take up so little room and ward away so many pests.

I grow: Thai, Genovese, osmin purple, and lemon basil.

Image 69 - Some of the marigolds in my hellstrip. I'm a bit of a fanatic now that I've discovered how well these work and how effortless they are to grow.

Image 70 - (Top) Edible nasturtium leaves. (Bottom) A happy bee feeding on a nasturtium flower in my garden.

Image 71 - (Top) Thai basil. (Bottom) Genovese basil.

THE IMPORTANCE OF CROP ROTATION

Crop rotation is simply the act of swapping what you grow in one spot to a different spot the following year.

You might wonder:

Why would I do that? If I got a great harvest and everything worked out last year, why shouldn't I do the exact same thing this year? After all, if it isn't broken, I shouldn't fix it.

In theory, that makes sense. All of your plants' conditions seemed to be met. But the answer to why you should rotate your crops actually sits below the surface of the soil.

When you plant a crop, that season, the roots of that crop are going to do everything that they can to suck up every vitamin and mineral that they need to offer you their maximum bounty. What happens when you plant the same thing there next year? Well, that area is already starting out pretty deficient in all the vitamins and minerals those plants needs to fuel themselves a second time around. By rotating your crops, you start to give the soil a break from those exact requirements, and it has time to start replenishing through your compost, fertilizer, or other organic materials that break down there.

The next crop you plant there might even benefit the soil by giving *back* some of what was taken, as is the case with a lot of legumes. They fix nitrogen back into the soil. Then, the next year, whatever gets planted there has a whole new supply of it to access.

You might have a dilemma if your garden is small, though. You might be thinking, *What if I only have one four-foot raised bed on my patio?*

Good news, you can still rotate your crops. You can always divide your bed into quadrants and rotate everything to the next square the following year. If you planted your tomatoes in corner #1, plant them in corner #2 next year, and the third corner the third year, and so on.

In a raised bed, you can do this by rows, shifting each row in one direction year after year. Or you can have some fun with it and scatter things around, moving them every year.

In addition to restoring nutrients, crop rotation can disrupt pest and disease life cycles, thereby increasing the amount of your harvest and decreasing the amount of your struggles. Pair that with homemade compost and those miracle pest deterrents I just mentioned in the last section, and you're likely to have yourself a healthy garden year after year that yields surprisingly large and more consistent harvests.

Wouldn't that be nice?

Rotating crops also helps a little with weed control, water retention, overall soil improvement, and fertility. It also cuts back on the need for those toxic and pricey chemical pesticides and store-bought fertilizers.

You can rotate crops via your own method, or, if you have *four* garden spaces or quadrants of a single space, on the next page is an example of a four-year crop rotation plan. Please note that you do not have to plant all of these things together, as some will clash (see *Planting Pals & Enemies*), but the lists show you what type of crops you can rotate.

Garden Bed 1

(Legumes. These fix nitrogen into your soil and replenish it after what has been extracted by other plants)

- Peas
- Beans
- Lentils
- Chickpeas
- Soybeans
- Clover
- Alfalfa
- Black-eyed peas

Garden Bed 2

(Root Veggies and Allums)

- Onions
- Chives
- Radishes
- Parsnips
- Leeks
- Turnips
- Shallots
- Carrots
- Beets
- Sweet potatoes

Garden Bed 3

(Fruiting Vegetables. These are heavy feeders)

- Tomatoes
- Peppers
- Okra
- Sweet corn
- Watermelons
- Honeydew
- Cantaloupe
- Pumpkin
- Squash
- Eggplant
- Cucumbers

Garden Bed 4

(Light Feeders. Since the soil has been somewhat drained by the heavy feeders, these can survive, and even thrive, on a lesser amount of nitrogen and nutrients)

- Lettuce
- Spinach
- Cabbage
- Brussels sprouts
- Broccoli
- Cauliflower
- Arugula
- Kale
- Swiss chard

If you keep your plants grouped like in this example, every year when you move the items from Garden Bed 1 into the second bed, and so forth, you will be ensuring a better balance of nutrients throughout the years.

COVER CROPS

Instead of leaving the garden bed or in-ground garden fallow in the fall and winter, many people choose to plant a cover crop once their plants have been harvested.

A cover crop is a non-cash crop planted in a garden bed or over a swath of land in an effort to cover the soil for cash crops later (as opposed to being harvested). Legumes and cover crops can improve the soil and reduce erosion by shielding the dirt from the impact of heavy rains and winds, therefore acting as organic armor to topsoil and causing a reduction in sediment loss. It does all of this while suppressing weeds, increasing the organic matter in your soil, *and* aiding with moisture retention and drainage. As added perks, they can also offer a beneficial habitat for pollinators and wildlife, as well as combat soil compaction.

Are cover crops absolutely necessary for your home garden? *Not really*. Although they certainly do help, especially if your soil is poor to start with.

One I highly recommend is crimson clover, like the Tommy James and the Shondells' song. The bees love it, it fixes a lot of nitrogen to the soil, it stays nice and short, lasts deeper into the cold months, and when you're getting ready to plant your garden again, you can till it up and leave the remnants right in the dirt It'll break down into effortless green compost.

TRELLISES AND SUPPORTS

Some plants need structure to grow properly.

For example, when most people think of a tomato plant, they picture it as a tiny little tree with leafy green boughs. But tomato plants are actually vines. Most cannot grow a stem thick enough or strong enough to stand on its own while holding the massive weight of its fruit. If allowed to grow wild in nature, most will even sprawl sideways along the ground like a watermelon or pumpkin vine.

To keep them from snapping in half or miring their fruit in the dirt, we have to provide them with structure. This can come in the form of a trellis, cage, teepee, strings, sticks, poles, cattle wire... You name it.

We are going to go over some of the most common methods of providing structure and protection for your plants, ranging from store-bought options to free things you can use around your home.

Continuing with our tomato example, the majority of gardeners will go out and buy their tomatoes a tomato cage. This is a wire support, often conical in shape, that goes around the plant like a tiny prison, and its spikes go into the ground. As the tomato vine grows, it can be massaged up through the center. Then, the metal supports hold the draped weight of the growing fruit.

Cages are great because they aren't terribly expensive, they come in different sizes, they're stackable at the end of the season, and they can be reused for quite a few years.

Plants that are ideal for a cage trellis include:

- Peas
- Tomatoes
- Peppers
- Eggplant
- Peonies

The tripod trellis, or teepee trellis, is made of several supports pressed into the ground and tied at the top, like a teepee. Then, twine or wire can be wrapped around it to secure the structure and give the plant more rungs to climb.

Plants that are ideal for a tripod/teepee trellis include:

- Cucumbers
- Peas
- Pole Beans

Image 72 - (Top) Cage trellis with tomatoes. (Bottom) Teepee trellis with cucumbers climbing it.

185

A-Frame and Arch Trellises are extremely versatile and strong. They can be bought and assembled or they can be as simple as two panels of cattle wire, fastened together at the top.

These types of trellises are ideal for:

- Watermelons
- Cantaloupe
- Honeydew
- Winter squash
- Banana melons
- Loofah
- Pole beans
- Cucumbers
- Tomatoes
- Squash

Image 73 - (Top) An A-frame trellis with tomatoes. (Bottom) An arch trellis with cucumbers.

186

Image 74 - My sugar snap peas using a length of spare rabbit wire as its trellis.

187

There are string trellises as well. These are extremely inexpensive, readily available, and can even be made at home if you've got the time and inclination. Basically, it is a string grid that can be fastened to any type of support in your garden. The plants send out their grabby little tendrils and latch on. Then, they wind their way up and through as the growing season continues on. As their fruit becomes more engorged and weighty, the string support helps the plant hold it until it's time for you to harvest.

Image 75 - A cucumber plant using its tendrils to latch on to my string trellis like a clenched fist.

Thick tree branch cuttings, bamboo poles, and coated metal garden supports are other great ways to keep your tomatoes and eggplants upright when they are starting to bear fruit.

To save money and go with a more natural look, I take my one or two-inch-thick, pruned branches from my fruit trees and shove them into the dirt. Then I use twine, plant tape, Velcro, or yarn to tie my tomato and eggplants to them loosely.

String trellises are great for tomatoes and eggplant, too, if they're all planted in a row. This is where a sort of children's swing-set frame is made out of supports with a horizontal crossbeam four or more feet above your row of veggies. Then, where swings would normally hang, you tie twine or yarn and let long lengths dangle down toward the base of your plant. Then, when you fasten the loose end to the stem beneath your plant's first set of branches, you are able to wind the string around the

plant to tighten the slack. As the plant grows, you keep winding in a circular motion around the stem until the twine is taut. This reinforces the main stem of the vine and keeps it upright as it begins to fruit.

189

MAGIC POWDERS

There are a few more tips and tricks that I, as a professional mistake-maker, am more than happy to bestow upon you to ensure that you have all of the knowledge you need to turn your small swath of soil into a great big harvest. These have to do with some magic powders that are either already at your home or easily found at any big box store.

Y'all ready for this?

Cue the *Jock Jams* CD! (Just kidding).

Elemental Sulfur - Sulfur is an acidifier. When planting your peppers (whether they're blazing hot habaneros or sweet green bell peppers), add some powdered sulfur in the hole first. Later in the season, your peppers will take off. Just make sure you don't use this near a plant that prefers alkaline soil, or it might suffer.

Note: In the big box stores, this often comes in bags of yellow or white granules labeled "soil acidifier," often with a photo of blue hydrangeas on the bag.

This can also be used as a top dressing mid-season, too. Just go slow so you don't *over*-acidify your soil either.

190

Powdered Milk - Ahhh, just like the kind mama used to make... *if you were like me and you grew up dirt-poor.*

But seriously... to get ahead of diseases caused by a lack of calcium in your soil, this is a fantastic trick to boost your transplanted seedlings and rocket your tomato plants to the moon.

Just like the sulfur, this works best if you pour a handful of it in the hole before placing your plant in it. As you water your plants through the season, the water turns the powder into liquid milk, and your plant has access to a gentle-on-the-roots form of calcium.

Epsom Salt - A bag of this stuff is likely sitting somewhere in your bathroom right now. It is fabulous for use in the garden, though, because it has a lot of magnesium and sulfur. Dissolve some in hot water, funnel it into a spray bottle, and spritz the yellow, blooming flowers of your tomatoes, eggplants, cucumbers, and melons. You

can also top dress those plants as well for more blooms and better flavor.

Just make sure not to overdo it, as too much salt can hurt your garden veggies, too.

Coffee Grounds - I've mentioned this one a few times already, but I wanted to include it here, too, as it is a magic powder. Just like with the sulfur or powdered milk, plant spent or fresh coffee grounds in the hole of any plant that enjoys a slightly acidic soil (like tomatoes and peppers) before transplanting your seedling.

Image 76 - Add your magic powder of choice to the hole before transplanting your seedlings. Or top-dress them throughout the season!

192

BASIC GARDEN PROBLEMS AND SOLUTIONS

In my opinion, half of gardening is about problem-solving and finding easy solutions that don't cost a ton of time or money. Every year seems to be something different, whether it's a freak hurricane snapping your stalks, a spotted lanternfly invasion, an all-out war with groundhogs, a pack of hungry deer, an uncharacteristically scorching summer heatwave, or any other number of problems.

As a gardener, you're likely going to have some casualties on the way, whether they are completely out of your control, or whether you are fully responsible. This is where we have to practice the patience and forgiveness we spoke about earlier in this book.

Fortunately for you, I have made a ton of costly mistakes, and I am here to share some easy fixes and preventative measures, as well as point out some things to keep an eye out for in your garden.

But first, a public service announcement...

A PSA ABOUT PESTICIDES AND SPRAYS

These are causing *irreparable* damage to the Earth right now. Full stop.

Not only are many companies like *Round-Up* paying out *billions* in lawsuits for using toxic chemicals that cause cancer in their users, but many pollinator species are dwindling or near extinction because of their widespread use. They pose deadly health risks to humans and animals, like respiratory issues, developmental issues, birth defects, eye and neurological issues, and more.

They damage ecosystems and dwindle pollinator populations of bees and butterflies. Firefly populations are waning, and eighteen species of them in North America are at risk of extinction. Over half of the fourteen native bumblebee species are in decline. The orange-and-black monarch butterfly has seen an 80% decline in the last twenty years. Dragonflies and odonata are being rapidly poisoned by the runoff from chemically-treated lawns. Ladybug and ground beetle populations are also in decline.

This is, in large part, due to the ease of access and the lack of knowledge about all of the sprays available at the big box and lawn and garden stores. They're cheap to get, promise the moon, and unless you have a lot of time and your magnifying readers handy, you aren't going to realize how bad these are for the environment.

You truly do not need them.

What you need are the insects they are killing, all in the name of having a uniform patch of grass sandwiched between your concrete sidewalk and residence.

Don't spray them. Don't risk your health, your dog's health, or the health of thirsty dragonflies or the adorable little lightning bugs taking a nap in your grass.

INSECTS

Insects are the bane of many a gardener's existence. I know I'm personally a bit frustrated with my constant war on spotted lanternflies once they arrived in Connecticut and started murdering all of my gorgeous grapevines with the slowest of deaths. But bugs don't have to break you as a gardener.

Here are a few problematic bugs and tricks to rid yourself of them without hosing all of your edibles down with toxic chemicals. I will not go into all of them, as that could fill another book entirely, but here are some of the common ones you might encounter in your garden.

A note about killing pests: I personally am a big animal fanatic. I was a wildlife rehabilitator in Louisiana with injured raccoons and opossums for years, and I've had a massive amount of exotic pets and wild animals over the decades. I want to state that I personally am not advocating for the taking of tiny lives en masse in this chapter. It is my general belief that living creatures deserve to exist, even if they are a nuisance to me. I personally spend more of my time and effort planting things like marigolds that *ward*

195

the pests away in the first place as opposed to killing them. You, however, may feel differently, and I do not judge you for that in the slightest. (I *will* judge you if you are hosing your lawn down with weed killers and/or pesticides now that you know what kind of damage they do, though).

I have decided to share some tips and tricks about pest control that I have learned about through gardening groups and books that work for a lot of people, even though many of these techniques I don't actually do in my own garden. I want to state that upfront because I do not want to come across as encouraging cruelty, but I also want to give you solutions that don't involve hosing your own food and beneficial pollinators down with toxic chemicals.

I'd bet that you're eating vegetables because they're healthy, so let's keep them that way by leaving those commercial sprays, gooey fly strips, and indiscriminate blue bug zappers out of the garden so that you can zone in on those tiny foes and not hurt everything else, yourself included.

As a cancer survivor, I avoid pesticides and chemical sprays like the plague. So much so that you can eat as you tour my garden. While I don't partake in fatal pest control methods, I'm not demonizing anyone who does, either. I know lots of people who treat their dogs or cats like royalty and wouldn't lose a single *wink* of sleep at night over smashing a spider. If that sounds like *you*, you might be able to glean some great non-pesticide tricks from this section. So let's get started!

APHIDS

Aphids are tiny, usually only 1/10 of an inch long. These pear-shaped terrors come in shades of green, brown, pink, or black. All have long antennae, and some have wings. Not only do they suck the stems, fruit, and leaves of your plants, causing them to yellow or pucker, but they can also transmit viral diseases like mosaic disease from plant to plant. They overwinter in an egg stage, and their presence is typically indicative of excessive fruit tree pruning or too much nitrogen fertilizer.

Image 77 - An aphid-infested leaf.

Their populations can also (ironically) become out of control if you are regularly using pesticides or weed sprays on your lawn or garden, as those tend to poison the aphid's natural predators.

All garden plants are vulnerable to these little menaces, unfortunately.

For light infestations, vigorously blast them with water every day or two in the early morning. For heavier infestations, using insecticidal soap and or neem oil every three days or so can help, though many people see mixed results.

My favorite methods for eradicating them, however, don't include chemicals. Instead, they involve luring natural predators so the circle of life can take care of them.

To manage infestations in a long-term way, plant trap crops, attract natural predators like ladybugs and avians, or grow companion plants that repel aphids.

Place bird feeders in or near your garden (preferably with a catch tray beneath. Birds are messy eaters, and their excess rain of seeds will have all sorts of random things sprouting up around the base of the feeder, such as sunflowers.)

Lure spiders, who prey on aphids and other harmful bugs, by planting camphor weed, goldenrod, or asters. Bring more fireflies in with tall ornamental grasses, especially native varieties. Entice predatory damselflies with alfalfa and wildflowers. Coax more assassin bugs with alfalfa, camphor weed, goldenrod, carrots, and oleander. All of these prey on those nuisance creatures.

CATERPILLARS

Many people are thrilled when these little guys show up in their garden, especially because they are growing milkweed, violets, or butterfly weed to help feed them and give them a nursery where they can develop into butterflies and moths.

However, there can be times when these leaf eaters eat a little too much of the plants you're growing for you, thus stunting their growth or annihilating them altogether.

While there are insecticides on the market that will poison these little guys (like BT, *Bacillus Thuringiensis*) and keep them from becoming a problem for you, I obviously want to discourage you from doing that, as these caterpillars, butterflies, and moths are pollinators. Killing them with poisons ensures their populations keep dwindling and cuts back on your overall pollinators.

Instead, repel them from the plants you want to keep with peppermint soap. Or, as I mentioned with the aphids, move your suet or bird feeder nearby so that the circle of life can continue to work in your favor.

Image 78 - A monarch caterpillar on a milkweed leaf.

EARWIGS

You know them well. During your childhood, you probably had nightmares about them crawling into your ears at some point. These things can chomp away at your foliage and cause your plants to look like some old, tattered flag flapping in the wind.

You can repel them with the oils of citrus, lavender, eucalyptus, cinnamon, or peppermint. You can also plant repellent herbs near the affected plant to drive them away. Earwigs don't like dill, basil, rosemary, or lavender.

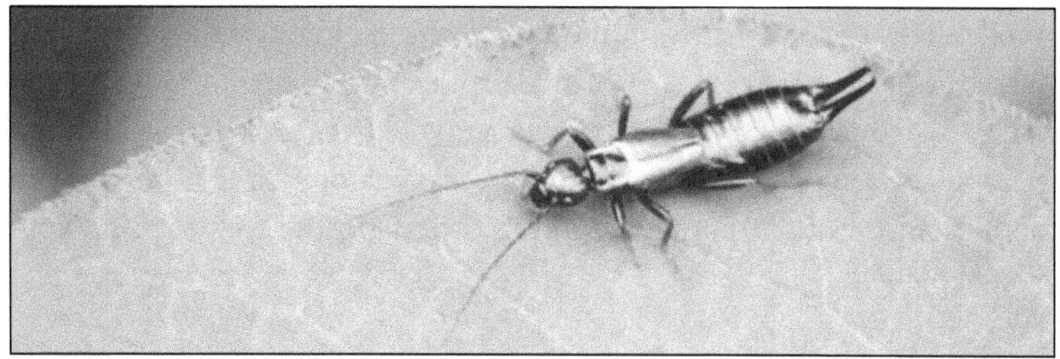
Image 79 - An earwig on a leaf.

If you'd rather a more permanent solution and have no qualms about killing insects en masse, you can rid yourself of these by burying a shallow container in the dirt near the affected plants. Then, fill it with soy sauce and a little bit of vegetable oil. They get attracted to the scent of the soy sauce and come out in droves. They slip inside, drink the soy sauce, and drown. It helps to put a lid on the container and cut a hole in the side near it to keep out rain or hose water, as the overflow can over-salt the soil near your plants.

SLUGS AND SNAILS

These can annihilate a lettuce bed. They love the cool, shady, damp conditions that lettuce thrives in.

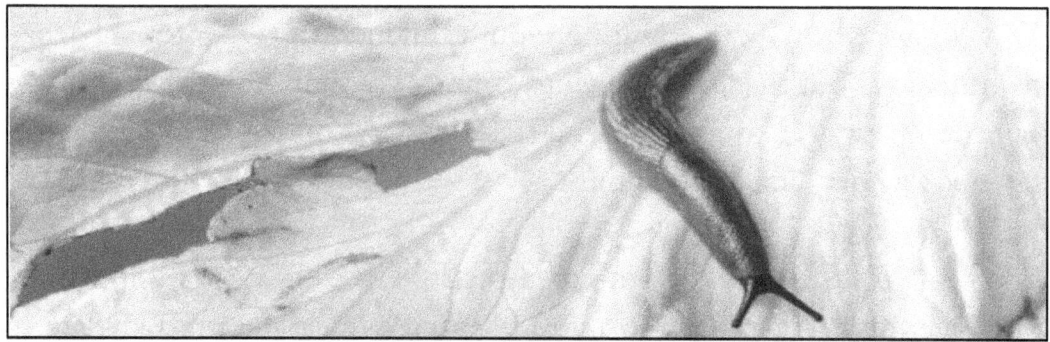
Image 80 - A slug on lettuce.

You can use a trap crop, like a hosta, to lure them away so that you can handpick them every night until your population is down.

Or you can do what I do. I keep them off my goodies with coffee grounds and smashed eggshells. They don't like the feel of the eggshell pieces under their soft foot, and they find coffee repulsive. Sprinkle some in your lettuce bed if you're starting to have problems with these adorable (but slimy) little things.

Alternatively, for a more lethal approach, you can use the same trick as the soy sauce under the section labeled *Earwigs*, only with baking yeast in water or beer. Either way, the slugs are attracted to the yeast. They slip in the liquid, get drunk, and drown.

POTATO BEETLES

These are convex yellow insects, roughly 1/3 of an inch long, marked with an orange head covering and black stripes. The yellow eggs are laid on the undersides of leaves. The grubs are fat with black and red spots and a black head. They are found all throughout the US and Canada.

Image 81 - The life cycle of the Colorado potato beetle.

201

As their name suggests, they particularly like to munch on potato plants. However, they also affect tomatoes and eggplants sometimes.

You can use a trap crop like the black nightshade to round them up. Then, once the trap crop is infested, you can yank and dispose of it in the trash (not your compost pile).

Another way to eradicate them is to sprinkle cornmeal or wheat bran meal around the infested crops. The meal expands inside of them after eating it, and they perish.

Or you can repel them like I do by planting catnip in a container nearby. (Then, during the late summer, I yank some of the catnip, dry it out, and bring baggies of it to my cat-owning friend's house as a free treat.)

SPOTTED LANTERNFLY

Prayer. That's about all you can do other than hit these things with a sandal or catch them by hand. They are extremely attracted to the invasive *tree of heaven,* so removing those from the area will help.

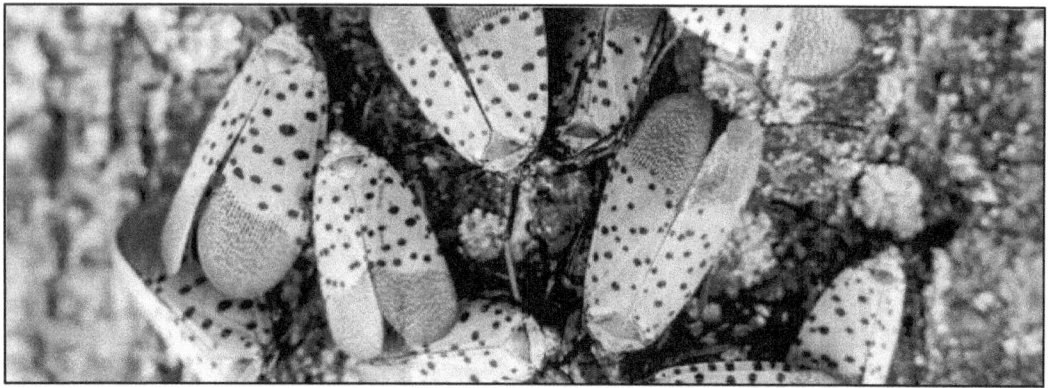

Image 82 - A tree trunk infested with spotted lanternflies.

They attack grapevines, maple trees, and anything that contains sap, slowly sucking the ever-loving life out of it.

Birds seem to hate the taste of them, and they breed like mad where I live because nothing seems to prey on them. Full disclosure: because these are such a problem locally and the state has asked us to eradicate them permanently whenever possible, I *have* been using lethal force with these to try to fight against the rapidly growing population. Not just for myself, but for the gardeners in my area. However, there are so many of them that resistance has been futile, thus far.

My contact information is in the back of this book. If you learn of a solution to this particular pest, I'd love to know about it.

SQUASH VINE BORERS

Vine borers love to feast on squash (hence their name), cucumbers, watermelons, and pumpkins. These things are awful.

One day, you are strolling through your idyllic garden, looking at the lush foliage in your pumpkin patch, wowed by the sheer size of your zucchinis. The next day, you come out to see the vines withering, dying from the end of yesterday's gorgeous new growth. The next day, they look even worse. You start to wonder if you're under-watering, or over-watering, or if the plant requires some sort of livestock sacrifice to the gardening Gods.

You would be wrong on all counts. You have just been paid a visit by a vine borer.

Squash vine borers are moths with dark front wings and transparent rear wings. These moths, however, are not really the problem. It is their eggs and caterpillars that are killing your precious plant baby. The moth lays its eggs on the stem of your vining plant. Once those hatch, brown-headed caterpillars march along the stem and chomp down, burrowing inside where they tunnel to feed on the plant from the inside.

Image 83 - A squash vine borer moth.

When you inspect your withering vining plant, you'll see debris that looks like sawdust outside a tiny hole. If you were to split that vine right there, you'd see one (or more) caterpillars wedged inside, their bellies full. By this point, unfortunately, it is usually too late to save the plant.

There have been, on occasion, people who have split their vine, picked out the caterpillars with a needle, patched it, and then used a hypodermic needle to inject BT into the vine and saved it.

I've never been able to successfully perform plant surgery like that. Instead, I sometimes wrap the first few inches of the stems of my vines with a cloth ACE bandage, or I wrap a few inches of aluminum foil around them, especially near the root. The vine borer moths can't lay their eggs inside in the early summer months, and by the time your vine has grown a lot longer, they are done laying their eggs.

204

TOMATO HORNWORM

The tomato hornworm is a green and white caterpillar, three to five inches in length. It gets its name from the protruding horn on its rear end. Don't worry, though. It can't sting you. These eventually turn into a moth, but while they are in this caterpillar stage, they can absolutely lay waste to a bed of tomatoes, sometimes overnight. They can also attack dill, peppers, eggplants, and potatoes, too.

You can handpick these little fellas early, if you see them.

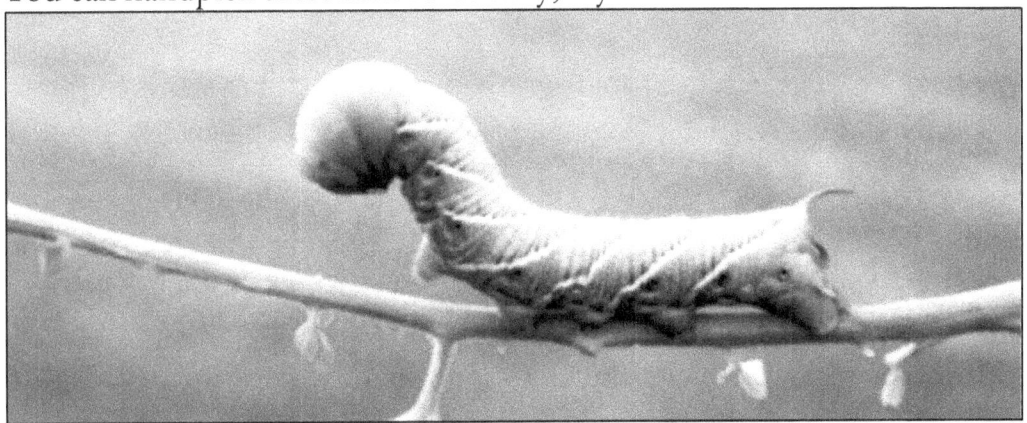

Image 84 - A tomato hornworm on a tomato vine.

If you're looking for a sacrificial or trap crop, I recommend dill. They love it, and as long as you go by that planting companion guide, they have lots of great garden buddies.

You can also repel them with a hot pepper spray (finely chop hot peppers, mix them with hot water, and then strain the peppers out so that they don't clog your sprayer).

Allegedly, they don't like neem oil either.

205

WASPS

Even though I think creatures (great and small) deserve to live, I also am in the camp of people who think *wasps just... suck.*

While they are still technically pollinators, these guys are like the bullies of the garden. One wasp sting could put someone off gardening for a lifetime.

So what should you do when they have taken up residence in your garden by building a nest?

If you are looking for a fatal solution, you can place a bowl of syrup out. They drown in the attractant.

But, if you're like me and you'd rather just ward them away, you can use repellents. You can grow citronella, basil, or mint. You can use peppermint oil on cotton balls or clip dryer sheets near the nest. They dislike the smell of all of these things. Having them around usually makes the wasps set up nests elsewhere near other food sources.

Image 85 - Wasps on a wasp nest.

206

DISEASES

There are a few plant diseases and viruses that might plague the garden you've worked so hard on. Just like the insects, there are many I am not touching on for brevity, but I want to give you some information and helpful tips about some of the ones you are most likely to encounter in your own personal veggie garden.

ANTHRACNOSE

Anthracnose is a group of fungal diseases that can occur in a vast array of garden plants. It usually presents as black sunken spots, often with light colored rings around each. This canker overwinters on branches of infected plants and trees. In the summer, these fungal infections commonly lead to your leaves dropping or browning, a reduction in the plant's overall vigor, and sometimes even spots on the actual fruits and veggies themselves.

It is important that any branches or foliage that you suspect of having anthracnose go in the trash can, not in your compost. These love steadily moist conditions (like your compost bin) and can make your problem tenfold next year if not disposed of properly.

To avoid anthracnose in the first place, I recommend using a healthy dose

207

of compost, which cuts down on fungal diseases as a whole (as I mentioned in the chapter on *Compost*). Using mulch also helps because it decreases the amount of dirt being splashed back up onto the stem and leaves of the plant when watering or during heavy rains. Yearly crop rotation and changing your watering schedule also help. Watering before noon is a great way to keep your plants from staying wet overnight, which is the ideal condition for most of these diseases to grow.

Plants that are especially susceptible to anthracnose are snap beans, lima beans, blackberries, cucumbers, pumpkins, mint, peppers, raspberries, squash, and watermelons.

To combat anthracnose, you can make a solution of one ounce of hydrogen peroxide plus two cups of water and mist the affected plants once a week with it. If this doesn't work, you may have to use a copper or sulfur-based fungicide. As I've mentioned before, though, these chemicals should be a last resort in my opinion.

BLIGHT

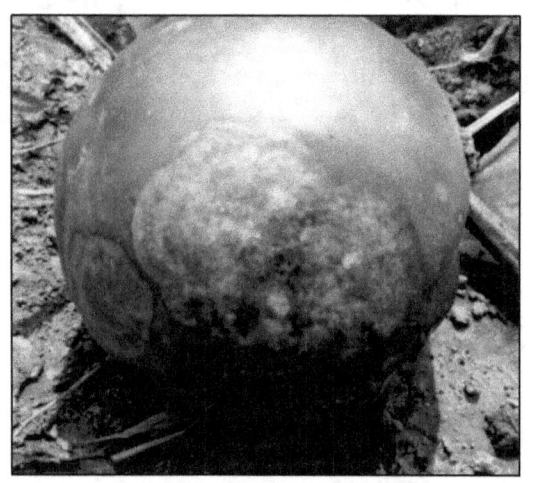

Blight is a bacterial disease that also likes wet, humid conditions to thrive and spread. It presents as large brown splotches on your leaves, often surrounded by red or yellow halos.

Blight is more likely to affect your tomatoes, beans, and peas.

Trim or pull affected plants and discard them in the trash. Make sure to disinfect your clippers afterward and not touch any of your other healthy plants after touching these without washing your hands first to keep this from spreading.

I have had some luck treating this with a diluted baking soda and water spray on the leaves and stems, along with crop rotation and compost.

Copper fungicide is another solution if you don't mind using chemicals in your garden.

BLOSSOM END ROT

Blossom end rot is a disease usually caused by a calcium deficiency, nitrogen overload, a heavy rainstorm, or improper watering during fruit formation.

This one is, fortunately, not very serious to the overall health of the rest of your garden.

It presents as black or puckered portions of your peppers and tomatoes.

If you have had heavy rains or have been over-watering, there isn't much you can do beyond dialing back to watering only when the plant *really* needs it and waiting for the plant to bounce back on its own.

If it isn't from too much water, check your soil with your soil meter. If it's

209

below a pH of 6, you might be lacking in calcium. An easy fix for this is top or side-dressing with some powdered milk or adding a little bit of limestone to the soil near the plant.

LEAF RUST

Leaf rust or orange rust is a fungal disease that tends to attack berry plants (like blueberries, strawberries, raspberries, etc) and fruit trees (like cherry and apple) more than anything.

It presents as spots of rust on the leaves.

This is a tricky one. If you start to see this, your plants will need to be pruned with the excess disposed of in the trash (again, not the compost pile), and then the remaining healthy leaves will need to be treated with either a hydrogen peroxide solution or a liquid copper fungicide.

POWDERY MILDEW

Powdery mildew is such a pain. It is a fungal disease that often goes unnoticed (at least in my garden) until it has spread like wildfire. It presents like a fine white powder on the leaves, sort of like powdered sugar.

Many plants are susceptible to this, including pumpkins, squash, watermelons, apples, blackberries, cherries, peas, raspberries, and strawberries. It usually comes from (you guessed it) the leaves staying too wet for too long. This can happen because of a rainstorm, high humidity, or watering too late in the day.

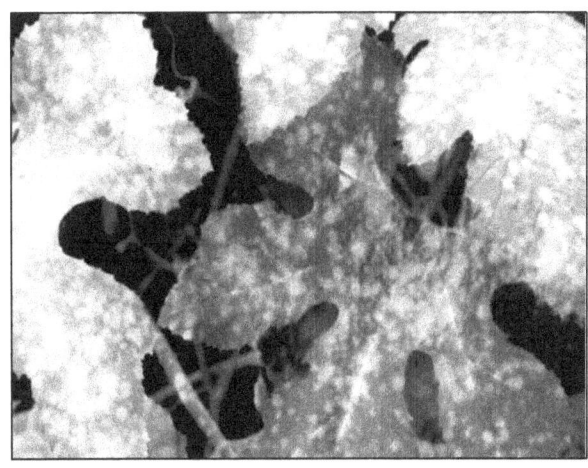

Prevent it by watering the roots (not the leaves) of your plants in the morning so the leaves all have a chance to fully dry before the sun goes down. Using drip irrigation or a canvas soaker hose also helps because you can water your plants without hosing down those leaves.

You can't control the rain, though. What should you do if your plants are already suffering from this stuff? Pruning to improve air circulation helps. Your plants might be too close together for their liking. You can also spray a diluted baking soda and water mixture, neem oil, or milk spray right on the leaves. The milk spray usually works like a charm for my plants.

SUNSCALD

Sunscald happens when the greens of your tomatoes are exposed to the sun for too long, sort of like a sunburn. It can also affect apple and cherry trees, too. This is usually caused by over-pruning or leaf loss and is easily fixed by dialing back on the pruning or giving it time.

WATERING LIKE A PRO

I mentioned this before, but just in case you skipped the chapter, I feel like this is worth repeating. When irrigating your plants, it is important to water the *roots*, not the *foliage*. Most people see gardeners on TV and in magazines showering their plants while waving at a neighbor with reckless abandon. That is a great way, when done at the wrong time of day, or too frequently, or after a series of rainstorms, to give your plants a whole host of fungal infections and diseases.

If you want to set your plants up for hands-free daily watering, get a couple of drip or soaker hoses at your local garden store along with a watering timer. When used together, your plants can get morning water without having their leaves drenched, all while you're fast asleep or on vacation. Drip irrigation is the easiest way to water without constantly having to battle fungal infections.

That said, I still water my plants by hand because I have a lot of plants with *different* watering needs (I grow a little of everything, and my watermelons and berries need about twice the water as my tomatoes and figs).

212

I also water by hand because it gives me another excuse to walk through my garden every couple of days and inspect everything. This is the best way to catch small issues before they become *massive* ones. I walk through and inspect for insect infestations as well as fungal issues. I look for animal damage or ripe fruits.

PERFECT PICKING

It is crucial to pick fruits when they are ripe, preferably with clean scissors or pruning shears.

Picking when ripe may sound like a no-brainer, but you'd be amazed at how many people procrastinate this and end up with smaller yields because of it. Picking keeps the plant from expending so much extra unnecessary energy in its quest to continue to ripen those particular fruits. Instead, it can focus on growing *new* ones, thus increasing your fruit and veggie hauls *exponentially*.

There are downsides to leaving your fruits and veggies on the plant too long. One is that your over-ripe fruits are bleeding your plants dry of the energy that could be spent creating new goodies.

Another downside is that the ripe or overripe fruits lure in hungry animals who also want to share your bounty. By waiting too long to pick those strawberries, you might realize the local squirrel has beaten you to the punch by sampling half of every one.

Another con is that leaving these things on the vine too long signals to the plant or fruit tree, "Well, that's about all the food I can handle outta you. You can go ahead and close up shop for the season and go to seed now."

When they aren't focused on fruiting, your plants are focused on making seeds or runners to ensure that they will come back next year, or their species won't die out.

But you don't need a million strawberry runners. You need strawberries! So get to picking!

This goes for leafy things like basil, lettuce, and kale, too.

There is one exception to this rule: your *tomatoes*. Those, you don't want to wait until they are ripe, if you can help it. You want to pick them just *before* they are red and ripe. They will usually turn from that lovely green to a blush color, almost a shade of light orange. This means they have entered into what is called the breaker stage. It means the fruit has everything it needs to ripen itself the rest of the way off the vine. So when you see those tomatoes turning anything but green, get the clippers out and trim them off. You can let them ripen right on your kitchen countertop (which they will do in one to four days, depending).

Picking when your tomatoes are in their breaker stage will make your plant breathe a (metaphorical) sigh of relief. It will say to itself, "Wow, that has taken a real weight off of me. Hmmm. What should I do now? Oh! I know! I'll make a bunch more!"

The next thing you know, you are giving away tomatoes to your neighbors and googling recipes for making homemade smoked salsa.

A WORD ABOUT GMOS

GMO stands for Genetically Modified Organism. This refers to any living thing whose DNA is lab-altered for improvement. These entered the scene in the 1990s as a way to solve food shortages and famine.

I know what I'm about to say will be considered a *hot-take* because of how demonized the term itself has become over the last decade or two. People have blamed just about every problem (from hangnails to cancer) on the existence of GMOs, but they also have saved countless lives, especially in areas of famine or with poor growing conditions. They're responsible for a higher efficiency in food production and food *security* across the globe.

Scientists have worked to create disease-resistant strains of seeds over the years that increase crop yields and reduce the need for spraying so many chemical pesticides that are hazardous to both pollinators *and* humans. Some GMOs have even been modified to *improve* their nutritional value. Golden rice is an example of this because of the higher vitamin A content. There are soybeans now that contain healthier oils, too, because of this process.

By producing GMOs with higher yields, stores have been able to lower the costs of these fruits and vegetables so that you can actually *afford* to eat healthier. Plus, some GMOs have been modified to require far less water so that they can be grown in drought-prone places like Africa and Australia.

Many studies have come out over the years about GMOs, some raising concerns over potential long-term effects, but as of the writing of this book, the consensus of almost all major scientific bodies (including the FDA and

World Health Organization) who have vigorous regulatory oversight over these organisms, have deemed the food from GMOs to be safe for consumption, if not *just as safe* as their non-modified counterparts. They also claim that the risks of consuming genetically modified foods, if any exist, still remain largely unproven.

There are also concerns that the patents on genetically modified seeds give corporations an unjust amount of control over our food supply and financially disadvantage local farmers. *This part might be true.*

The good news is that a lot of the GMO seeds out there aren't going to be in the packs you buy at your garden center. They are used on a much wider scale than personal gardens, which means you are mostly going to encounter GMO food in your local grocery store or at a restaurant, versus your patio garden.

If you are concerned about GMOs, keep an eye out for packs of seeds that say things like "100% Organic!" or "Heirloom" or "F1 Hybrid" or that have the Non-GMO Project Verified Logo on the pack. Most of the seeds that you will find in the store, by brands like Burpee, will already be GMO-free.

Table 1 - Rinsed veggies from my garden. Most of what is available to you is already organic and non-GMO.

216

VARMINTS

Just like with the insects, it is my stance that all animals deserve to live. After all, they are only in your garden because they're hungry, just like you. Then again, very few of us are going to put the time, effort, money, and space into gardening as a hobby just to feed it all to the local wildlife.

Here are a few of the animals you may encounter in your garden and some tips and tricks on what to do about them. Just like the insects, you are able to use lethal force to protect your garden if you're the kind of person who has that hunter instinct that people like me lack.

For all the rest of you, here are some ways to deter, repel, or humanely relocate the creatures plaguing your home-grown berry patch.

DEER

Deer are majestic and overall pretty harmless to humans, but they can lay your fruit trees and garden flowers to waste in a matter of minutes. Building

a fence really helps, but since they can leap and gallop over most chain link with ease, you might need to go a step beyond that to deter them from chomping away at everything you just grew.

The thing I've found that really helps is buying an inexpensive pack of bars of soap, drilling holes in

217

each, and looping some twine or string through. Then, I hang it wherever the deer seem to like to hang out. They hate the smell of the soap (especially *Dial* and *Irish Spring*), and they will often stay away.

I have also heard they hate the smell of garlic, so spraying garlic-water around your fence line should help, too.

DOGS

Dogs are usually easy to keep out with a solid fence, but if your dogs are diggers, like mine, you might need to reinforce the base, too. I had Jack Russells, and I ended up having to line the bottom of my raised beds with chicken wire, too, just to stop them from tunneling up into it like a gopher.

It is especially important to keep them out of your garden if you are growing grapes or muscadines. These are toxic to dogs. The last thing you want is for some to get jostled on their trellis and fall in your yard and sicken or kill your furry bestie (or your neighbor's dog).

Please, grow these vines responsibly if there are canines in the vicinity.

GOPHERS AKA GROUNDHOGS

These little creatures can be an absolute pain in a gardener's butt. And usually, if they're a problem for you, they are a problem for anyone

gardening within a mile or two of you, too. Their underground tunnel systems can spread up to two miles wide.

For these, I recommend catching them in humane traps and relocating them to a wooded area miles away.

If you have no intention of doing such a thing, there *are* methods of repelling them, too. They hate lavender, human hair, lemon balm, mint, sage, basil, rosemary, cayenne, castor oil, and chives, so covering any gopher entrance holes and planting or pouring any of these in or around it will help.

You can also plant a row of sacrificial crops, such as cabbage, around the perimeter of your fence so that they will eat those instead.

In my experience, rehoming the entire family of these things is usually the only way to ensure I get to eat the spoils of my hard work, though. They are ravenous, stubborn, and can fit in holes and crevices the same size as any rat, even though they can be the size of a dog.

219

OPOSSUMS

While these may munch on your corn or berries, I usually leave them be. Opossums do *not* carry rabies, they're immune to venom, and they will feast on snakes in your garden.

If you have dogs, they will keep your flea and tick population down, too.

Often, these little creatures do more *good* than harm. I never re-home these. If I catch one, I re-release it back in my yard's maple tree and let it do its thing.

If you are dead-set on keeping them away, though, you can try repelling them with garlic spray, hot pepper spray, or rags soaked in ammonia.

RACCOONS

These are similar to opossums in that scents like garlic, hot peppers, and ammonia seem to deter them. It is also important, if you have a prevalence of raccoons in your area, not to place your trash cans anywhere near your garden, as the smell of food or even carrion is what attracts these little bandits.

Note: Raccoons are a rabies vector species,

220

so I do not advise handling them. Check out your state's laws on raccoons to see if there are any rehabilitators who can take them if you catch one in a humane trap.

SKUNKS

These stinky little devils can be quite cute up close. They are very easy to catch in a humane trap. However, they are hard to re-home without your vehicle holding that unmistakable stench for a while.

If these are a problem in your garden, I recommend fortifying your fence's base a little and leaving them be if you are not (understandably) willing to relocate them

They also aren't fans of the smell of citrus peels or citrus oils, or ammonia-soaked rags, so you can easily use these as a deterrent as well.

SQUIRRELS

They're cute. They're cuddly looking. And they love to snack on the spoils of your garden.

Please exercise caution when re-homing these little guys, as they are a rabies vector species.

If you don't want to re-home them, you can boil

221

some Tabasco pepper and garlic powder, strain it, and add a little dish soap. Then, you will have a spray that, when applied liberally, will make these little guys think twice about rooting around in your garden. That said, I have had squirrels that basically still kept going like they were on the Hot Ones wing challenge, unbothered in the slightest.

I am a big fan of humanely trapping and re-releasing these cuties elsewhere.

HUMANE TRAPS

These humane traps (also cleverly called a *Havahart* trap) are a great way to catch and transport the pests in your garden and a solid investment, in my opinion. You can get these online for $15-$50, depending on the size of the critter you are going for, and so far, I've been using mine for years and have never had a single incident of an animal escaping on my way to re-release it elsewhere.

These traps can be baited and set fairly easily, often catching critters within hours as opposed to days. Use the fruit or veggies the animal seems to be gnawing on as bait and set the trap in or near your garden where you've seen signs of their activity or have witnessed their presence.

Some animals are not *technically* legal to re-release elsewhere. Get familiar with your state laws. Many do it anyway as the penalty is often a warning or small fine, but it's always best to search your local laws first and know your options. Sometimes, there are even licensed animal caregivers for things like raccoons that will come and get the creature and release it elsewhere after an evaluation.

Image 86 - One of the gophers I caught feasting on my pumpkins in my raised bed.

Be especially careful handling rabies vector species such as the squirrel and raccoon. One bite from one of these and you will end up having to have a whole bunch of painful rabies shots, whether you are infected or not.

I take the animals a few miles away, especially the gophers, because their tunnels can spread far underground. All they have to do is find one entrance back in a mile away and they will return to visit you in no time. I also always take mine to a large swath of woods where they are more likely to make a home in the wild than become the next gardener's problem.

223

WHERE ON EARTH ARE YOU GROWING?

HARDINESS ZONES

A hardiness zone or a grow zone is a geographic region that is defined by the colder side of its annual temperatures. These zones are used by gardeners and farmers to determine what perennial crops, flowers, trees, or ornamental grasses are likely to survive in that region.

The US Department of Agriculture (USDA) developed the most widely used system, which is illustrated on the back of most seed packs in your region. This USDA hardiness system divides the US into thirteen different zones based on ten-degree Fahrenheit segments, dividing *those* into "a" and "b" half zones as well.

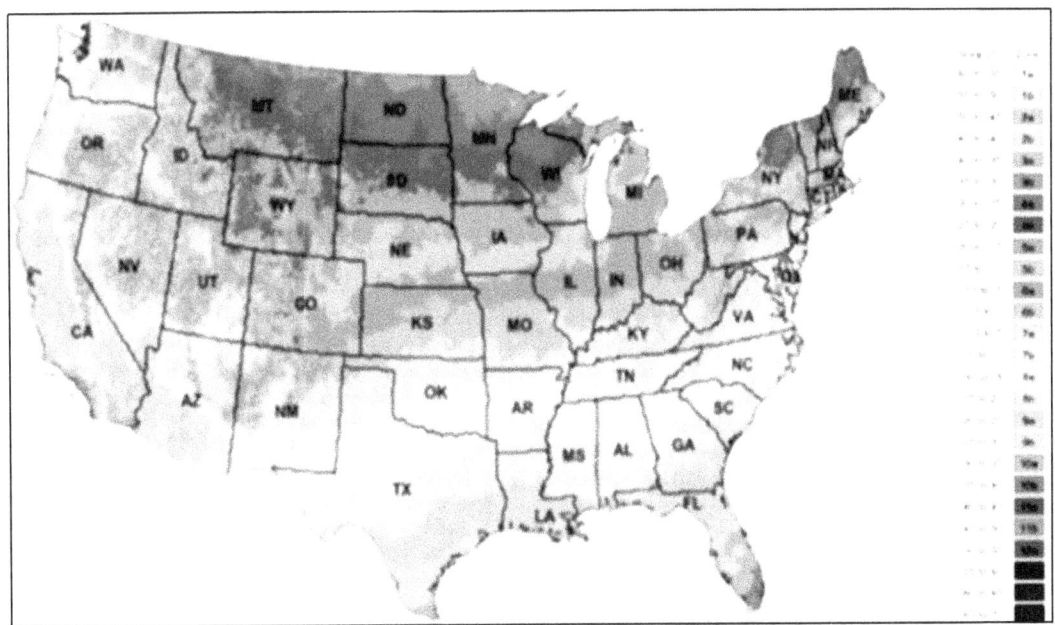

Image 87 - A map of US grow zones.

Why are these grow zones so important? Because it helps to know what plants will be an annual (meaning they will die off every year) or a perennial (meaning they will come back during a certain season every year) in your area. It will also help you decide what you can grow successfully and when to direct sow your seeds.

Here's an example. I used to live in southern Louisiana, which is grow zone 9 here in the United States. That meant a whole host of plants that would be considered an annual up in, say, Montana, would be perennial where I lived. It also meant I could grow some things, like bananas, year-round, which would normally not survive the harsh winters up north.

Conversely, when I moved to southern Connecticut, I moved to zone 7a. The downside of this was a shorter growing season, and plants like my canna lilies would have to be overwintered inside in a pot, or the rhizomes would have to be dried and stored for the winter and replanted again after the last frost in spring.

The upside? I could finally grow cherry trees in Connecticut. In Louisiana, while a cherry tree would produce plenty of foliage, there aren't enough chill hours in a year down there to get a cherry tree to produce fruit.

Chill hours are the number of hours during a tree's dormancy that the buds are exposed to temperatures between 32 and 45 degrees Fahrenheit. This period of cold is necessary for ensuring it will produce fruit come springtime. Not enough chill hours can lead to a lack of fruit, smaller fruit, and fewer blossoms.

Back to the grow zones, these are precisely why, in 2025, the US tariffs the president imposed raised the price of certain things for Americans (like coffee and bananas) substantially. Americans have to import the majority of those particular things from other countries.

According to the grow zones, most of the United States (save for some areas of Hawaii, Florida, Louisiana, and Georgia) doesn't contain climates suitable for growing tropicals such as bananas or coffee. Those thrive in places much closer to the equator, like Ecuador, Colombia, Brazil, Costa Rica, Guatemala, and Vietnam.

Knowing your grow zone will help you figure out what plants and trees are going to not just survive -- but *thrive* -- in your area.

OVERWINTERING PLANTS

Overwintering is the process of bringing something inside for the winter that would normally be considered an annual in your area. For example, I do this often with my pepper plants. Let me illustrate this with a hypothetical that I have actually done:

It's fall. The cold winds are coming with regularity. The leaves of the trees are going to start changing any day now. But my Tabasco pepper plant hasn't gotten the memo. It is simply *loaded* with peppers!

It seems like such a shame to call it quits when the plant is doing so well. That is one place where overwintering can come in handy! I find a nice deep pot and dig my Tabasco plant out of the ground and get it situated in the container. After a few days of letting it acclimate, I decided to bring it inside. After all, I don't want it to freeze and wither when this weekend's storm rolls in. Now, suddenly, I have this pepper plant inside that I can pluck fresh peppers off of for the weeks to come, squeezing a few extra weeks of juice out of planting it in the first place.

But that's not even the best part! Say I am able to keep that thing alive all winter in its pot by overwintering it inside. After the last frost of the following spring, I can drag that pot outside, re-plant the mature Tabasco plant, and start getting loads of new peppers out of that thing. The extra benefit to all this effort is that the plant is already mature when it goes back in the ground, so it can get right back to work instead of starting all over as you would from a seedling.

I once did this with a Tabasco plant for three years. Every year, the plant grew larger and put out exponentially more Tabascos. Right before my cross-country move, I ended up selling the plant (that I'd started from an inexpensive seed three years ago and eaten from for three seasons) for fifty dollars! I got a photo that fall from the buyer who was showing off the literal bucket of tiny peppers he'd just pulled off it.

Overwintering is great if you can get the hang of it. You're buying one seed or one plant and getting several years of goodies out of it! Also, if you know you're going to be overwintering it beforehand, you can always grow it in the container to start with, so you never have to traumatize those roots. Just keep feeding with compost and fertilizer, and you're good to go!

Image 88 - Some houseplants and cacti on my overwinter shelving racks in my old kitchen.

228

THE DOUBLE POT METHOD

This is another one of my handy-dandy tricks of the trade.

In the winter, I bring a lot of plants inside. I like the jungle feel, and I also like growing some things, like lemons, that can't survive the cold months outside. These grow *fantastic* in the summer outside but must come in when the temperatures drop.

In the first few years of gardening, I simply dragged these outside and lined them all up. They were a little bit of an eyesore since the pots were mismatched, but more than that, the wind often knocked them over. I broke a few branches each year off of my lemon trees, lime trees, and fig trees by toppling them. Then, I had a great idea (surely I didn't come up with something so clever, but for the life of me, I can't remember a source for this stroke of genius, so, for now, I'm going to pretend I invented it, even though I'm certain I did not.)

I call it the double pot method. This is super simple. All it requires is a few larger pots with drainage holes. Simply dig a wide hole, deep enough for your bigger pot. This is your outer pot. This outer pot stays in the ground year-round. You never take it out. It just acts like one of the cup holders in your car. The drainage holes ensure the rain and hose water go back into the soil instead of drowning your inner pot.

Once you have filled the loose dirt back in around your outer pot, you can simply put your inner pot (the container that holds the plant that comes inside every fall) inside. That's it! Now the plant is easy to rotate, so you don't get lopsided growth. The inner pot drains into the outer pot, which drains into the soil. The wind can't knock it over. And you can easily pull the inner pot out come autumn and bring your baby right inside. The outer pot sits vacant over winter, and then after the last frost passes, you can put your overwintered plant right back outside in its spot.

I did this for several plants this last year, and I won't be going back. It works like an absolute dream!

Image 89 - My figs and variegated pink lemon tree nestled in their outer pots in my yard.

THE LAST FROST

The last frost is the date in spring when the last freezing temperatures are expected to occur in your area. These dates vary widely. You can search your area's last frost online and get a loose estimate based on historical weather data, but remember, it is only an *estimate*. If you aren't sure if the last frost has passed, keep your seedlings mobile in flats or cups and keep an eye on the weather. If the temps dip too much, bring them back inside.

The last frost is a *super* important guideline for farmers and gardeners. Knowing this date upfront will also give you a good idea of when to start your seedlings indoors. Just be certain that your last frost has passed when you transplant. Jumping the gun, so to speak, can kill off every seedling you just worked hard to grow. This pretty much happens to all of us, and eventually, you will start getting wise to it.

I cannot tell you how many times I have spent February and March and part of April growing my seedlings inside, getting them nice and healthy, over-zealously transplanting them in the ground, and then watching in horror three weeks later as a snowstorm buries them under inches of white powder. It is a big setback when this happens, but it isn't the end of the world, either. You can always direct-sow more seeds or start more in flats indoors. It just reduces the amount of harvesting time by a couple of weeks.

Note: The date varies *significantly* depending on your elevation, location, and proximity to water. For example, the average last frost date for my area in southern Connecticut is roughly April 20th, while it's May 10th, a little closer to the middle of the same tiny state. I've also seen it snow in my area in early May, weeks after my last frost was said to be long gone.

ANNUALS VERSUS PERENNIALS

I mentioned these earlier in the section on hardiness zones. An annual is a plant that blooms, lasts for an entire grow season, and then dies. A perennial is a plant (usually a bulb or rhizome) that you plant once, and it comes back year after year, usually during a specific season.

These vary by your grow zone (or hardiness zone) because some plants can withstand certain low temperature thresholds. If the temperatures in your area never drop below that zone, what would be an annual elsewhere, blooms for multiple seasons where you are. This happens a lot in the southern states of North America. A lot of plants that are considered annuals here, where I am, in New England, were considered perennials where I was in southern Louisiana.

You might think to yourself, *Why should I bother with annual plants that I have to start from scratch every year when I could just put in a bunch of perennials once and call it a day forever?*

You certainly *can* do that. However, there are upsides and downsides to both types of plants. One major reason that gardeners don't do this is that a lot of the things they want to grow (like eggplants or tomatoes) are *only* annuals. Unless they overwinter inside successfully, these plants have to be restarted every season.

Perennials are great, but one notable downside is that they vigorously bloom for one short, gorgeous burst and then the flowers die back, and you are left with sad-looking foliage or sticks the rest of the year. Tulips and daffodils, for example, are the first stunning signal that spring has arrived.

232

These bloom for a couple of weeks, and then you are left with bladed leaves and foliage for the rest of the summer. If you blink, you will miss them sometimes. Irises and lilies are the same, only they usually bloom in the summertime. There are fall perennials, too. These are fun to pepper around your yard and garden to add some interest, but annuals are the flowers that typically are the "milkshake that brings all the bees to the yard."

Annual flowers (like cosmos, black-eyed Susans, marigolds, etc may only live for one year, but once they bloom, they are prolific and put out lots of pollen-filled flowers week after week that will attract a myriad of pollinator insects. They look great and draw in beneficial bugs the entire garden season. Many also vigorously reseed (like marigolds, nasturtiums, morning glories, and black-eyed Susans) at such a rate that they might as well be perennial because if you plant them once, you will usually have them coming back year after year.

A warning about morning glories: Three years ago, I bought a pack of purple morning glory seeds and planted them along a chain link fence. In two months, you couldn't even see the chain link anymore. It was just a sheer wall of flowers that bloomed every morning and enticed all the bees. I

233

didn't know that they reseed themselves to an almost invasive level. Three years later, I'm still picking them from my fences on the complete opposite side of my property.

Image 90 - My chain link wall of morning glories.

The good news is, plants like that couldn't be more low-maintenance. You simply have to allow them to grow wherever you want them for the years to come. The bad news is, if you decide in three years you hate the look of that type of flower, you're going to be cursing every time you have to pick them because they spring up like weeds and literally put a stranglehold on your trees and such.

Are there *edible* garden perennials? Yes!

Strawberries, asparagus, raspberries, blackberries, grapes, and kiwi (along with fruit trees) are great edible perennials. Plant these once, add compost yearly (except to the grapes. Those prefer nutrient-drained soil, actually.

Too much fertilizer will make them grow massive vines and very little fruit), and these edible perennials will come back year after year, each time bigger, stronger, and more productive than the year before.

Image 91 - (Top) A colander of strawberries. A single weekend spring haul from one bed. (Bottom) A closeup of a young strawberry growing.

235

BULBS, SEEDS, ROOTS, AND RHIZOMES

Bulbs, rhizomes, roots, and seeds are all different parts of a plant used for germination or reproduction. I want to explain the differences in these quickly in case you find yourself in the garden department looking for things to grow this year.

A seed is a small unit that contains an embryo and a food supply all of its own. When water penetrates the seed shell, the baby plant inside can germinate and start seeking warmth and sunlight above the soil's surface.

Most garden vegetables and annuals start out as a seed, including cucumbers, squash, pumpkins, lettuce, cosmos, zinnias, marigolds, etc.

A bulb is most commonly a teardrop-shaped organ of organic material (similar to an onion) where the plant stores its energy. Plants that reproduce through asexual methods typically store their energy in a bulb like this that can reproduce on its own.

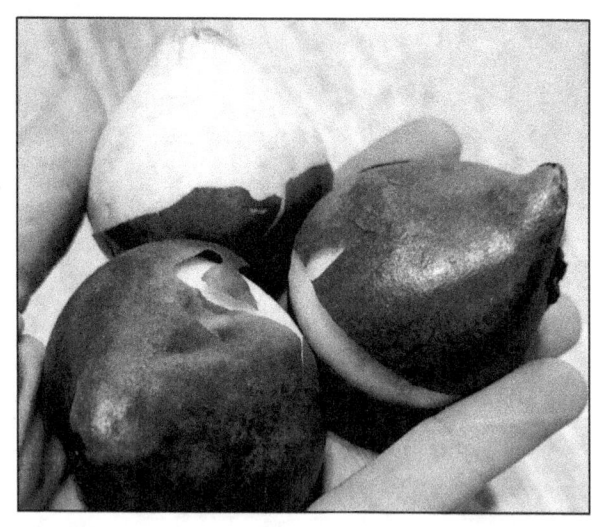

236

Gladiolus, daffodil, onions, and tulips are typically sold as bulbs.

A root is a fibrous system typically composed of a crown with tendrils. You might also hear the term bare-root being associated with these. Bare-root means that the soil-free cluster of roots has gone dormant for the season and is ready to be planted in your garden at the appropriate time.

Strawberry, grapes, roses, and asparagus roots are often sold this way.

A rhizome is a usually tuberous stem that grows horizontally underground, producing buds along its length. These are typically perennial plants. Some are also edible.

Common rhizomes are ginger, horseradish, irises, mint, rhubarb, and bamboo

THE SEASON FINALE

Well, your growing season has come to an end. The weather is getting chilly, and it is time to harvest the last of your crops. Now is your last chance to enjoy another round of cold-weather crops like lettuce and spinach. It is also the time to plant your cover crop, if any.

This is also the time to harvest seeds for next year or turn spent plants into compost.

Here are some neat tricks for the end of your garden season:

THE TRENCH

One neat trick that saves a lot of time for me in my in-ground bed is that I dig a deep trench (think of it like a shallow grave for someone as tall as Yao Ming or something). Then, I take all of my spent plants (save for any affected by blight, fungal infections, or insect infestations... those go in the trash), chop them up a little with some shears, and lay them all inside the

238

trench I dug. I layer in some shredded junk mail and tree leaves, cover it all with the dirt I just dug out, and water the lumped-up soil once really well.

Why do I bother with all of that?

Because in about an hour, I just vermicomposted all of my old plants at once without taking up any yard or composter space. This fattens up those hungry worms and leaves behind a whole garden row of nutrient-rich castings that will benefit my garden next year.

I just turned my dead plants into effortless fuel for next year's garden.

A note of warning on this method: make sure you don't throw any pieces of grapevines or blackberry canes in the trench, or you'll have a bunch of volunteers (plants that grow even though you didn't intend to plant them) sprouting up in that row the following year. I learned this the hard way the spring after I first did this. I kept walking around, going "This looks like a grapevine. I didn't plant grapes anywhere near here, though!"

That said, if you want to make a ton of free baby grapevines or blackberry bushes, dump these scraps in a container with dirt. Next year, you'll see a whole bunch of them springing up.

THE IDEAL WATERMELON

Knowing when your watermelons are ready can be tricky. When picking your watermelon, here are some notes on what to look for:

239

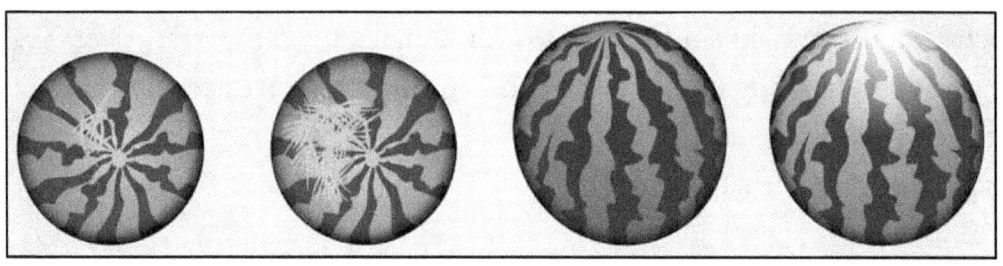

Smaller webbing = bland
Larger webbing = sweet and flavorful
Dark and dull = ripe and ready
Shiny = not ripe yet

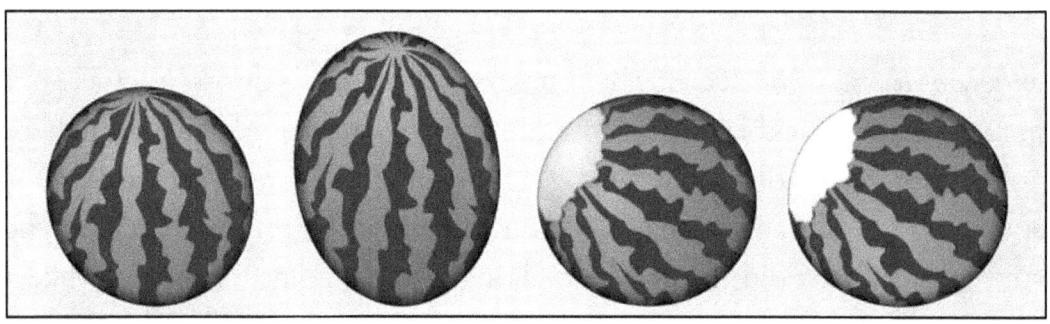

Uniform size and heavy = sweet
Elongated = watery
Orange field spot = full of flavor
White field spot = very little flavor

FREEZING, PICKLING, AND MAKING SAUCES

Did you overdo it with the tomatoes this year? Do you have enough jalapenos left to last you for the next three months of taco nights? Do you have enough concord grapes to make a jar of jam? What should you do when you have too much to eat that is ready right now?

240

I often give a lot of my excess away to friends and neighbors. After all, it's a nice treat to receive something fresh, free, and totally organic. But what if you want to enjoy this stuff through the winter?

You have lots of options:

- Turn those extra tomatoes into smoked salsa or mix in some basil and make some homemade marinara!

- Boil tomatoes and freeze your homemade tomato sauce in containers.

- Slice zucchini and yellow squash and freeze them in plastic freezer bags for a frozen veggie dinner side in the winter.

- Turn those buckets of banana peppers and jalapenos into jars of pickled peppers for sandwiches. Pickle those cucumbers or red onions, too!

- Turn those extra strawberries, figs, or blackberries into jam or preserves.

- Freeze excess broccoli, cauliflower, and Brussels sprouts in gallon bags.

- Turn that extra squash into chocolate zucchini bread.

- Learn how to ferment your own small batch wine with your green grapes!

Whatever you have fun making with the spoils of your garden will encourage you to grow more of that next year. Soon, you'll be gifting homemade treats and have a fully stocked food pantry!

HARVESTING SEEDS

Harvesting seeds is slightly different for each plant. Get to know each as the method is often super easy once you know what to do. Some are more obvious, such as a giant sunflower, while some are more subtle and hidden by a flower or in a pod, such as basil and tobacco. Pulling your plant *before* it has gone to seed means you are missing an opportunity to re-stock your seed organizer with free goodies. If something does well in my garden, I always harvest the seeds from it.

Image 92 - A bucket full of seeds from a row of giant sunflowers I grew. I roasted some as snacks and used the rest for birdseed the next year to thank the birds for keeping the aphids out of my garden.

242

Some seeds are easy to retrieve because they grow in a pod, like milkweed seeds. These are as simple as squeezing the pod, scooping out the fluff and seeds, and then separating the downy material (as I will show you in the Penny Trick coming up).

Other seeds can be trickier because they are encased in a layer of protective goo. A perfect example of this is tomato seeds.

Let's say I loved my black krim tomatoes this year, and they were *such* a favorite that I want to grow them again next year for sure. I get to work by cutting open one of my mature black krim tomatoes and scooping out the seeds. They're each covered in a gelatinous substance that doesn't really want to come off, even when I rinse them in a flour-sifting colander with tiny holes. So what do I do? *I really want to grow these again!*

Fermentation is the answer. This is a super simple process that takes very little effort. At the end of your season, when you pull in the last of those black krims, collect all the jelly-covered seeds into a small jar with a lid. Something like a baby food, pickle, or capers jar. Then, add in some water and put the lid on. Shake the jar vigorously. Then, let it sit for a week or so. This process is called fermentation.

In a week, open the lid and pour the contents of the jar into that flour-sifting colander with the tiny holes (so the seeds don't just slip right out into your sink and go down the drain). *Caution*: this liquid is going to reek to high heaven for a few minutes. It's going to smell like dog flatulence, so just... *prepare thyself*. With your sink sprayer, spray the seeds and watch all of the tomato bits and gelatinous muck slip down your drain, leaving you with nothing but a whole bunch of black krim seeds! Set them in front of a fan or on a piece of wax paper so that they dry *fully* and don't mold or

243

sprout. Then, save them in an envelope or baggie for spring. It's *that* easy.

Image 93 - (Left) a jar of fermenting melon seeds. (Right) Strained seeds after a rinse.

Don't forget to label them. I can't *begin* to tell you how many times I have thought, *Oh, I'll remember what this is,* and then forgot immediately. *Spoiler alert*, months later, you'll most likely *not* remember. Not to mention, most seeds in the same families look pretty much identical.

Image 94 - Spacemaster cucumber seeds next to a ramekin of banana melon seeds. They're identical.

Image 95 - Pumpkins are one of the easiest things in the garden to harvest seeds from.

245

Image 96 - I put my pumpkin seeds in a container of water, shake vigorously, and then strain.

You can also space damp seeds out on a strip of single-ply toilet paper and make your own seed tapes.

A seed tape is a great way to get perfectly-spaced seedlings in neat rows in your garden.

Just plant the seed-studded toilet paper (seed facing up) at the right depth and *boom*, you've got yourself a nice row of black krims (or cucumbers or anything else) for free!

Plastic baggies or jewelry baggies are a great way to save your seeds, too.

These come in various sizes and are great for separating some of the seeds you harvested to give to neighbors and friends as well!

Image 97 - I pluck my marigold seed pods in the fall and end up with bags of them!

248

Image 98 - My basil has gone to seed as evidenced by the flowery stems.

THE PENNY TRICK

Here's a fun trick to separate all the chaff and down from your seeds. I call it the Penny Trick:

Step one: Collect your seed pod. To illustrate this, I have chosen a milkweed pod.

Step two: Over a Ziploc baggie, pop open your seed pod, or for something like a morning glory, roll it with pressure between your fingers until it cracks and seeds start to fall out. In the case of this milkweed pod, I have extracted the downy material and the seeds together with my fingers and placed them inside the baggie.

Step three: Add a few pennies to the bag (any coin will work, really.) Then, seal it.

250

Step four: Shake vigorously. Let the pennies clang around. These will give the seeds the right amount of abuse to break free from their chaff or downy material. This is the material that protects them and/or ensures the seeds cling to things along the ground in the wild, so they stand a chance of making their way into the soil.

Step five: After quite a bit of vigorous shaking, you now have seeds that are separated from the chaff or down. You can either pierce a seed-sized hole in the bottom of the baggie and let them fall out into your container, or you can reach in and pinch out the unnecessary material (the white, furry down in this instance) and throw it away, leaving behind a baggie full of seeds!

Note: Milkweed seeds need to be cold-stratified, meaning they either need to experience a real winter through the baggie or a mock-winter in the refrigerator for a few weeks to simulate winter. Some seeds are programmed to wait until "winter" has passed before ever springing to life. So if you were to bring these inside and keep them warm all winter and then plant them in the spring, they likely wouldn't sprout. You'd think that you ended up with a bunch of duds when in reality, the following year, you'll end up with a milkweed bonanza! So make sure to do a little research on whether your seeds need to feel some of the cold before they go in the ground.

LEAVE THE LEAVES!

This is a section that is very near and dear to me. It concerns one of my biggest pet peeves in the world: fall leaf clean-up.

I get truly upset when I see people bag their leaves and put them on the side of the road for landfill pickups. Unless they are diseased or carry major fungal infections...

Leaves and grass don't belong in the landfill. Full stop.

Where *do* they belong? *In your yard*. Either mulched, piled, or in your compost. Every time you send these to the dump, you're doing harm to the planet, harm to pollinator species and small animals, *and* robbing your lawn and/or garden of free fertilizer!

Honestly, sending your leaves to the landfill is bad for the environment for several reasons. For one, leaves are often buried under trash and left to decompose anaerobically, a method that releases methane, which is a greenhouse gas that is both potent and a large contributor to climate change and the earth's steadily rising temperatures. If the leaves are sent to the dump in plastic bags on top of it, this adds to the plastic waste and environmental pollution. It also contributes to landfill overcrowding, which is a widespread problem worldwide. After all, we have floating garbage islands because we are running out of places for our insane amounts of waste.

Conversely, leaves that are either mulched or left to decompose naturally actually feed and enrich the soil, provide excellent nutrients like nitrogen, and give winter homes to an unfathomable amount of small animals and billions of pollinators.

If you belong to a rigid Homeowner's Association that doesn't *allow* leaf litter, consider mulching leaves with your lawnmower instead. The piles will be gone, and the fragments will settle down into the grass and dirt and decompose, giving your lawn and garden more nutrients to come back lush and full the following year. This is why you can go hiking in the forest even though no one ever rakes there. Leaves are organic and break down over a short period of time, leaving the ground beneath better than it was before.

A whole ecosystem exists beneath your leaf litter. Creatures (such as small mammals, birds, amphibians, and pollinator insects) need this habitat for camouflage and cover, breeding, foraging, feeding, insulation, over-wintering, and general survival. Leaf decomposition also attracts earthworms that help recycle your unwanted leaves and grass clippings back into nutrient-rich castings for your soil.

Leaving your leaves in your yard doesn't *just* save time and effort. It can prevent injuries, too. Nearly 40,000 raking-related injuries are reported *annually*... just in the United States alone!

Skip the pulled muscles and the greenhouse gases. Do something great for the environment and *leave those leaves!* Or *compost them* to feed your raised beds, containers, and in-ground gardens.

You can also dig a trench and bury the leaves, as I mentioned earlier in this chapter, to concentrate your worm castings in a specific garden area.

If you *must* clean your lawn of leaves for aesthetics, or for a picky HOA, consider *mulching* them with your lawnmower instead. Make sure to leave the bag off and just let those fragments settle into the grass. The wind and rain will carry the detritus down to the soil, where it will become food for a lot of living organisms instead.

The trees, animals, pollinators, and earthworms will thank you for it. As will your garden the following year when it is full of the nutrient-rich compost you made!

Image 99 - Leaf litter is some of the best brown compost there is!

255

CONCLUSION

Almost nothing in this world brings me more joy than being out in my garden, soaking up the sunshine, and playing in the dirt.

One of the best perks to this activity is seeing your efforts spread throughout the area around you to your neighbors, friends, or relatives. It puts a smile on my face every time someone says my garden inspired them to start their own or jump back into it after a hiatus. It warms me every time I get a thank you letter from co-workers for the bags of fresh vegetables they all got to take home and make fresh meals with. And, of course, it is a joy every time I bite into a sub sandwich I made at home with organic food that was still growing on the vine that morning.

It doesn't get any fresher than that!

It is my hope that in compiling my years of notes, documenting my mistakes, telling you about my tribulations, and sharing lots of research with you that I have given you the confidence to start (or improve upon) your compact garden.

Start small. Stay patient. And know that you will never have a 100% success rate, no matter how many years you decide to stick with it. That's part of the excitement and thrill of it all.

I once had a gardener tell me the secret to his huge, lush garden. He said, "Erica, there's no magic to it. I just grow a lot. I start a lot of seeds and hope that no one realizes how much I kill in a given season." This surprised me because, from the sidewalk, his garden looked like some sort of professional nursery, like he should be giving seminars on growing. But the more we walked around on his garden tour and the more stories he told me about gophers or aphids or leaf rust claiming some of his plants, the more I realized that gardening isn't a perfect science. There are many factors beyond our control that can make us lose a plant (or many).

But man, is it ever a fun endeavor. It is my heart's deepest desire that I can spread my love of gardening to many more people while I'm on this planet. It's rewarding. It's fun. It's good for your health. It's surprisingly cheap. And it is never the same from one year to the next.

Thank you for buying this book and for entrusting me with your time. I am no master gardener. I still make mistakes. Still, I have been able to see the astounding improvement of my *own* garden over the years, and I want that for you, too.

I wish you years of happy gardening ahead!

-Erica Summers

257

CONTACT ME

Have questions or spotted lanternfly eradication suggestions?
Want to tell me about how this book affected *your* compact garden?

Feel free to reach out to me at: trixiefairdale@gmail.com
I'd love to hear from you!

FUNKY PLANTS I THINK YOU SHOULD TRY

Lemon Cucumber - These small, round, yellow cucumbers are prolific and taste amazing. They're the perfect size for a sandwich, and they're so crisp and crunchy. These are an absolute staple in my garden. These do well on a trellis.

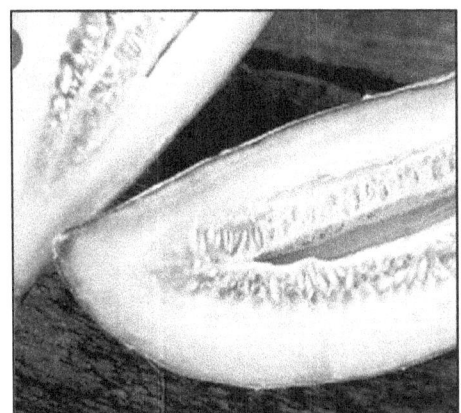

Banana Melons - These long green melons are so fun and funky. When you cut them open, they taste like a cucumber and a cantaloupe had a baby together. The fruit is refreshing and goes great in cold water (make it just like how you would make cucumber water!) These require a trellis.

Bush Baby Watermelons - These are space savers and don't take up a ton of room. The watermelons they put out are just a little bigger than a softball, so they're more "personal-sized" than "family-sized" but wow, they pack a lot of great flavor!

259

Black Krim, Black Beauty, or Indigo Rose Tomatoes - I *love* growing black tomatoes in my garden to mix it up a little. I find black krims especially juicy and flavorful. I love tomatoes, and these are probably my all-time favorite kind to grow.

Loofah - Yep! You can grow your very own loofah sponge. I bet you didn't even know those were plants! They grow great on hanging archways and trellises.

Okra - The flowers on okra plants are so pretty. I often have people walk up to my fence while I'm watering and say, "What is that gorgeous yellow flower right there? I've never seen anything like it!" I always have to resist the urge to laugh because it is an okra plant, and the hibiscus-like flower turns into the okra eventually. Plus, bees love them!

Black Nebula Carrots - These purple carrots are sweet, huge, crunchy, colorful, and so much fun to grow. They are bright purple and really liven up a salad. I put them in homemade chicken soup once, and it turned the whole soup purple!

260

ACKNOWLEDGMENTS

I would first like to thank my sister and writing partner, Heather Wohl, for encouraging me to write this. You are my best friend and, as always, this simply wouldn't exist without your support and guidance. Thank you for always encouraging me.

I want to thank Dave Sikora for not only supporting my gardening habit but also for getting involved in it with me. Nothing makes me quite as giddy as looking out my office window and seeing you perusing the garden with interest and excitement after you get home from a long day at work. Watching you grow your first plants from seeds was one of the best moments of my whole planting journey. Here's to many more years of growing our own fresh food together!

I want to thank my father, Jim Kane, for imparting his gardening knowledge upon me through the years. Your passion for it is a huge part of what made me want to try it in the first place. Thank you, also, for allowing me to use photos of some of your container plants and your beautiful greenhouse for use in this book.

I wish to thank Roseanne, Juanita, and Anne Marie, three of the coolest neighbors a gardening gal could ever ask for. Watching you work on your gardens, talking with you, troubleshooting issues, and sharing the spoils of my harvests always reminds me why I find this hobby so inspiring.

To Ralph Stevens for his gardening questions over the years. Some of the things that you've asked me actually inspired whole sections of this book.

For Lindsey Rodriguez. Thank you for all of your unwavering support of *all* of my writing endeavors.

For Joey Green, Jerry Baker, Dick Raymond, Nancy Bubel, and Jeff Ball, whose non-fiction books saved my garden *numerous* times over the years. You taught me *so* much about gardening, and you never even met me.

Image 100 - Every year is a new, blank slate!

262

ABOUT THE AUTHOR

Erica Summers is an award-winning author and filmmaker, an artist, a film industry grip, and a cancer survivor. Born in Wyoming, she spent most of her adult life in central Florida and southern Louisiana. She now lives by the beach in southern Connecticut. When she isn't writing or reading, she is tending her vegetable garden, solving escape room puzzles, singing karaoke, or kayak fishing.

She has published a wide variety of award-winning horror and fantasy genre fiction and children's books. She also writes contemporary romance under the pen name Odessa Alba and hilarious cozy mysteries under the pen name Trixie Fairdale.

MORE BY RUSTY OGRE PUBLISHING

From Ashes
BOOK ONE OF THE ILLUMINATOR SAGA
Heather Wohl

Desdemona In Embers
BOOK TWO OF THE ILLUMINATOR SAGA
Heather Wohl

The Undead Queen
BOOK THREE OF THE ILLUMINATOR SAGA
Heather Wohl

Call of the Wyl
A DESTORIAN STANDALONE MYSTERY NOVELLA
Heather Wohl & Aurora Alba

Rumspringa

Odessa Alba

A NEW ENGLAND BILLIONAIRES BOOK

Tangled Heirs

Odessa Alba

A NEW ENGLAND BILLIONAIRES BOOK

The Ugly Sweater PARTY

a Holidale Romance Novel

AURORA ALBA & ODESSA ALBA

The Great Timbers

James A. Kane

The Choice is Yours

Yakshar's Lost Treasure

By Rowen Sikora
& Erica Summers

The Choice is Yours
The Wishing Gem
By Rowen & E. Sikora
with Erica Summers

The Choice is Yours
The Australian Outback
By Heather Watt
& Erica Summers

The Choice is Yours
Escape From Sugarland!
By Heather Watt
& Erica Summers

The Choice is Yours
The Branson Ranch
By Heather Watt
& Erica Summers

INDEX

RUSTY OGRE PUBLISHING